# Magnetism and Superconductivity in Iron-based Superconductors as Probed by Nuclear Magnetic Resonance

Franziska Hammerath

# Magnetism and Superconductivity in Iron-based Superconductors as Probed by Nuclear Magnetic Resonance

Foreword by Prof. Dr. Bernd Büchner

RESEARCH

Franziska Hammerath
Dresden, Germany

Vollständiger Abdruck der von der Fakultät für Mathematik und Naturwissenschaften der Technischen Universität Dresden zur Erlangung des akademischen Grades eines Doktors der Naturwissenschaften (Dr. rer. nat.) genehmigten Dissertation.

ISBN 978-3-8348-2422-6 ISBN 978-3-8348-2423-3 (eBook)
DOI 10.1007/978-3-8348-2423-3

The Deutsche Nationalbibliothek lists this publication in the Deutsche Nationalbibliografie; detailed bibliographic data are available in the Internet at http://dnb.d-nb.de.

Springer Spektrum


*Cover design:* KünkelLopka GmbH, Heidelberg

Printed on acid-free paper

Springer Spektrum is a brand of Springer DE. Springer DE is part of Springer Science+Business Media.
www.springer-spektrum.de

Für Madeleine

# Foreword

Since iron and its compounds are often ferromagnets - a notation that originates from the latin word “ferrum” for iron - superconductivity is not the first property to check for in an iron-based compound. It was therefore even more surprising that the superconducting transition temperature could be raised up to 55 K shortly after Kamihara and co-workers reported the discovery of superconductivity in the so-called iron pnictides in 2008. In fact, this is the highest transition temperature ever, seconded only by the cuprates, which were found in 1986.

Similar to these, the observation of another family of high temperature superconductors led to an enormous activity within the research community, resulting in thousands of papers presenting surprising and also promising new results. For example, high critical current densities and high critical fields raise the hope that these materials could be well used for applications. On the other hand, the mechanism of superconductivity is still not clear. While a simple phonon-mediated Cooper pairing seems to be excluded by the high transition temperatures, it is not yet established if antiferromagnetic spin fluctuations can indeed mediate the pairing, or if they are just a remnant of a nearby spin density phase. Also, the symmetry of the order parameter seems to change from compound to compound. While most of the experimental results are in line with the theoretically proposed $s_{+-}$-symmetry, there is substantial evidence for $d$-wave symmetry from other experiments. Even the possibility of $p$-wave superconductivity has been proposed, but with contradictory results on the experimental side.

All these intriguing questions are addressed in the PhD thesis at hand. In this thesis, Franziska Hammerath has used the experimental technique Nuclear Magnetic Resonance (NMR) in order to investigate several different compounds of superconducting iron pnictides. NMR has been invented already 60 years ago, and contributed to the field of superconductivity from the beginning on. To name but a few examples, the peculiar temperature dependence of the spin-lattice relaxation rate observed by Hebel and Slichter in the superconducting state of aluminum supported the theory of superconductivity by Bardeen, Cooper and Schrieffer, and the constant Knight shift in $Sr_2RuO_4$ in the superconducting state provided evidence for triplet superconductivity in this compound. Furthermore, its local character and the sensitivity to low frequency magnetic excitations render it a perfect tool to investigate also the normal state properties of a new superconducting family, such as the iron pnictides.

This thesis comprises a coherent summary of the basics of NMR as well as a brief description of the fundamental properties of the iron pnictides. With this at hand, the interested reader can immerse into the enlightening results of Nuclear Magnetic Resonance on iron-based superconductors, gaining a deeper insight also into the details of this powerful experimental tool as well as a better understanding of the iron pnictides as a new bearer of hope in the research of high temperature superconductivity.

Prof. Dr. Bernd Büchner
Director of the Institute of Solid State Research
IFW Dresden

# Contents

**Foreword** vii

**List of Figures** xiii

**List of Tables** xv

**1 Introduction** 1

**2 Basic Principles of NMR** 3
2.1 Isolated Nuclear Spins in a Magnetic Field . . . . . 4
2.1.1 A Single Nuclear Spin . . . . . 4
2.1.2 Many Nuclear Spins . . . . . 5
2.2 Making Life Interesting: Interactions . . . . . 6
2.2.1 Interactions Between Nuclear Spins - Line Shape I . . . . . 7
2.2.2 Magnetic Hyperfine Interactions . . . . . 8
2.2.3 Electric Quadrupole Interactions . . . . . 11
2.2.4 Summary . . . . . 16
2.3 Dynamic Processes: Relaxation . . . . . 17
2.3.1 How to Measure Nuclear Magnetism? . . . . . 20
2.3.2 Spin-Lattice Relaxation . . . . . 22
2.3.2.1 Sources of Spin-Lattice Relaxation . . . . . 22
2.3.2.2 Inversion Recovery Method to Measure $T_1^{-1}$ . . . . . 24
2.3.2.3 Spin-Lattice Relaxation Functions . . . . . 24
2.3.2.4 Korringa Relation . . . . . 26
2.3.3 Spin-Spin Relaxation . . . . . 27
2.4 Line Shape II . . . . . 29

**3 NMR in the Superconducting State** 31
3.1 Knight Shift . . . . . 31
3.1.1 Spin Shift . . . . . 31
3.1.2 Orbital Shift and Spin-Orbit Scattering . . . . . 32
3.1.3 Diamagnetic Shielding . . . . . 33
3.2 Spin-Lattice Relaxation Rate . . . . . 33
3.2.1 BCS Coherence Factors . . . . . 33
3.2.2 BCS Spin-Lattice Relaxation Rate . . . . . 35
3.2.2.1 Comparison to Ultrasonic Attenuation . . . . . 37
3.2.3 Absence of the Hebel-Slichter Peak . . . . . 38
3.2.4 What to Expect for Pnictides? . . . . . 39

**4 Iron-based Superconductors** **41**
4.1 General Overview . . . 41
4.1.1 Crystal Structure and Electronic Structure . . . 41
4.1.2 Ground States and Phase Diagrams . . . 42
4.1.3 Symmetry of the Superconducting Order Parameter . . . 44
4.2 Basic properties of $LaO_{1-x}F_xFeAs$ . . . 47
4.3 Basic properties of LiFeAs . . . 50

**5 Experimental Setup** **53**
5.1 Magnets and Cryogenics . . . 53
5.2 Electronic Measurement Equipment . . . 54
5.2.1 Probes . . . 54
5.2.2 Main Elements of the Electronics . . . 55
5.3 Samples . . . 56
5.3.1 $LaO_{1-x}F_xFeAs$ . . . 56
5.3.2 LiFeAs . . . 58
5.4 Problems and Improvements . . . 60
5.4.1 Temperature Control . . . 60
5.4.2 Sample Probe for the 16 T Field Sweep Magnet . . . 62
5.4.3 Sample Heating . . . 64

**6 NMR on $LaO_{1-x}F_xFeAs$ in the Normal State** **67**
6.1 Knight Shift - Static Susceptibility . . . 67
6.1.1 Optimally-Doped $LaO_{0.9}F_{0.1}FeAs$ . . . 67
6.1.2 $LaO_{0.95}F_{0.05}FeAs$ . . . 74
6.1.3 Discussion . . . 75
6.1.4 Summary . . . 76
6.2 Spin-Lattice Relaxation Rate - Dynamics . . . 76
6.2.1 Optimally-Doped $LaO_{0.9}F_{0.1}FeAs$ . . . 77
6.2.2 Underdoped $LaO_{0.95}F_{0.05}FeAs$ . . . 81
6.2.3 Overall Doping Dependence - Role of Spin Fluctuations . . . 82
6.2.3.1 Qualitative Discussion . . . 82
6.2.3.2 Quantitative Discussion . . . 85
6.2.4 Summary . . . 91
6.3 Korringa Relation . . . 92
6.3.1 Summary . . . 96

**7 NMR and NQR on $LaO_{1-x}F_xFeAs$ in the Superconducting State** **97**
7.1 Spin-Lattice Relaxation Rate for $x \geq 0.05$ . . . 97
7.2 The Effect of 'Smart' Deficiencies: $LaO_{0.9}F_{0.1}FeAs_{1-\delta}$ . . . 102
7.3 Summary . . . 108

**8 NMR and NQR on LiFeAs** **111**
8.1 $^{75}$As-NQR . . . . . 111
8.2 $^{75}$As-NMR . . . . . 115
8.2.1 Static NMR Properties - Linewidth and Knight Shift . . . . . 115
8.2.2 Dynamic Properties - Spin-Lattice Relaxation Rate . . . . . 120
8.3 Summary of $^{75}$As NQR and NMR Results . . . . . 124
8.4 $^{7}$Li-NMR . . . . . 125
8.5 Summary of $^{7}$Li NMR Results . . . . . 130

**9 Conclusions** **131**

**A Appendix** **135**
A.1 NMR Powder Spectra . . . . . 135
A.2 Calculation of Spin-Lattice Relaxation Functions . . . . . 139
A.3 Stretched Exponential Relaxation Function . . . . . 143
A.4 Spin Diffusion . . . . . 147

**Bibliography** **151**

**Acknowledgement** **171**

# List of Figures


2.1 Level splitting due to quadrupole effects in zero field. . . . . . . . . . . . . 14
2.2 Level splitting due to first-order quadrupole effects in a strong magnetic field. . . . . . . . . . . . . . . . . . . . . . . . . . . . . . . . . . . . 15
2.3 Summary of all possible interactions and the resulting NMR spectra . . . . 17
2.4 Effective field in the rotating frame (non-resonant case) . . . . . . . . . . . 19
2.5 Effective field in the rotating frame at resonance . . . . . . . . . . . . . . . . 19
2.6 Relaxation and precession of the nuclear magnetization . . . . . . . . . . . 20
2.7 Fourier transformation of a rectangular pulse . . . . . . . . . . . . . . . . . 21
2.8 Hahn spin echo pulse sequence . . . . . . . . . . . . . . . . . . . . . . . . 22
2.9 Inversion recovery method . . . . . . . . . . . . . . . . . . . . . . . . . . 25
2.10 Relaxation rates as function of the correlation time $\tau_c$ . . . . . . . . . . . 29

4.1 Crystal structure of iron-based superconductors . . . . . . . . . . . . . . . 42
4.2 Electronic structure of LaOFeAs . . . . . . . . . . . . . . . . . . . . . . . 43
4.3 Phase diagrams of cuprates and pnictides . . . . . . . . . . . . . . . . . . . 45
4.4 Early NMR results in the superconducting state of pnictides . . . . . . . . 46
4.5 Momentum-dependent superconducting gap in $Ba_{1-x}K_xFe_2As_2$ . . . . . . . 46
4.6 Phase diagram of $LaO_{1-x}F_xFeAs$ . . . . . . . . . . . . . . . . . . . . . . 48
4.7 Susceptibility of LaOFeAs . . . . . . . . . . . . . . . . . . . . . . . . . . 49
4.8 ARPES results on LaOFeAs . . . . . . . . . . . . . . . . . . . . . . . . . 50
4.9 ARPES results on LiFeAs . . . . . . . . . . . . . . . . . . . . . . . . . . 52

5.1 Resonant circuits . . . . . . . . . . . . . . . . . . . . . . . . . . . . . . . 54
5.2 Experimental setup . . . . . . . . . . . . . . . . . . . . . . . . . . . . . . 55
5.3 Field sweep spectra of aligned and misaligned $LaO_{0.9}F_{0.1}FeAs$ powder samples . . . . . . . . . . . . . . . . . . . . . . . . . . . . . . . . . . 57
5.4 $^{75}$As NQR resonance line of LiFeAs at room temperature . . . . . . . . . . 59
5.5 $^{7}$Li NMR spectra at 200 K in $H = 4.4994$ T . . . . . . . . . . . . . . . . . 60
5.6 $^{75}(T_1T)^{-1}$ of LaOFeAs upon cooling with nitrogen and helium . . . . . . 61
5.7 16 T sample probe before and after improvements . . . . . . . . . . . . . . 63
5.8 Effect of sample heating on the $^{75}$As NQR frequency of LiFeAs . . . . . . . 65

6.1 $^{75}$As-NMR resonance line of *ab*-aligned $LaO_{0.9}F_{0.1}FeAs$ . . . . . . . . . . . 68
6.2 Knight shift vs macroscopic susceptibility in $LaO_{0.9}F_{0.1}FeAs$ . . . . . . . . 69
6.3 Pseudogap fit on $^{75}K_{ab}$ for $LaO_{0.9}F_{0.1}FeAs$ . . . . . . . . . . . . . . . . . 70
6.4 Hyperfine coupling constants for $LaO_{0.9}F_{0.1}FeAs$ . . . . . . . . . . . . . . . 71
6.5 $^{75}$As Knight shift of $LaO_{0.95}F_{0.05}FeAs$ . . . . . . . . . . . . . . . . . . . . 74
6.6 Scaling of $^{75}K_{ab}$ and $\chi(T)$ for $LaO_{0.95}F_{0.05}FeAs$ . . . . . . . . . . . . . . . 75
6.7 $^{75}$As $(T_1T)^{-1}$ of $LaO_{0.9}F_{0.1}FeAs$ . . . . . . . . . . . . . . . . . . . . . . 78

6.8 $^{57}$Fe, $^{75}$As, $^{139}$La, and $^{19}$F $(T_1T)^{-1}$ in $LaO_{0.9}F_{0.1}FeAs$ . . . 79
6.9 Scaling of $^{57}$Fe, $^{75}$As, $^{139}$La, and $^{19}$F $(T_1T)^{-1}$ in $LaO_{0.9}F_{0.1}FeAs$ . . . 80
6.10 Dynamic and static NMR properties of $LaO_{0.95}F_{0.05}FeAs$ and $LaO_{0.9}F_{0.1}FeAs$ . . . 81
6.11 Spin dynamics of $LaO_{1-x}F_xFeAs$ and $Ba(Fe_{1-x}Co_x)_2As_2$ . . . 82
6.12 Phase diagrams of $LaO_{1-x}F_xFeAs$ and $Ba(Fe_{1-x}Co_x)_2As_2$ . . . 84
6.13 Normalized $(T_1T)^{-1}$ versus $T/T_c$ for different superconductors . . . 86
6.14 SCR (Curie-Weiss) fits for undoped (underdoped) $LaO_{1-x}F_xFeAs$ . . . 87
6.15 BPP fits for underdoped $LaO_{1-x}F_xFeAs$ . . . 89
6.16 Korringa relation for $LaO_{0.9}F_{0.1}FeAs$ . . . 92
6.17 Korringa relation for $LaO_{0.95}F_{0.05}FeAs$ . . . 93
6.18 $(T_1T)^{-1}$ vs $K$ in $LaO_{0.89}F_{0.11}FeAs$ . . . 95

7.1 $^{75}$As NMR $T_1^{-1}$ of $LaO_{1-x}F_xFeAs$ for $0.04 \leq x \leq 0.1$ . . . 98
7.2 $^{75}$As NQR $T_1^{-1}$ of $LaO_{1-x}F_xFeAs$ for $0.05 \leq x \leq 0.1$ . . . 99
7.3 $^{75}$As NMR $T_1^{-1}$ of $LaO_{0.9}F_{0.1}FeAs$ in the superconducting state . . . 101
7.4 Resistivity of $LaO_{0.9}F_{0.1}FeAs_{1-\delta}$ and $LaO_{0.9}F_{0.1}FeAs$. . . . 102
7.5 $^{75}$As NMR powder spectrum of $LaO_{0.9}F_{0.1}FeAs_{1-\delta}$ . . . 103
7.6 Recovery curves of $LaO_{0.9}F_{0.1}FeAs_{1-\delta}$ at different temperatures . . . 104
7.7 $^{75}$As spin-lattice relaxation rate for $LaO_{0.9}F_{0.1}FeAs_{1-\delta}$ . . . 105
7.8 Comparison of the $^{75}$As spin-lattice relaxation rate for $LaO_{0.9}F_{0.1}FeAs_{1-\delta}$ with pure and Co-doped $LaFeAsO_{0.89}F_{0.11}$ . . . 107

8.1 $^{75}$As NQR spectra of LiFeAs . . . 112
8.2 $^{75}$As NQR quadrupole frequency of LiFeAs . . . 113
8.3 $^{75}$As NQR $(T_1T)^{-1}$ of LiFeAs . . . 114
8.4 FWHM of $^{75}$As NQR and NMR spectra of LiFeAs . . . 116
8.5 $^{75}$As NMR Knight shift of LiFeAs at low temperature . . . 117
8.6 $^{75}$As NMR Knight shift of LiFeAs in the whole temperature range . . . 119
8.7 $^{75}$As NMR $(T_1T)^{-1}$ of all LiFeAs single crystals . . . 121
8.8 $^{75}$As NQR and NMR $(T_1T)^{-1}$ of LiFeAs samples S1 and S2 . . . 122
8.9 $^{31}$P NMR $(T_1T)^{-1}$ of $La_{0.87}Ca_{0.13}OFeP$ . . . 123
8.10 $^{75}$As NQR and NMR $T_1^{-1}$ of LiFeAs sample S3 . . . 123
8.11 $^{7}$Li NMR spectra of LiFeAs . . . 126
8.12 FWHM of $^{7}$Li NMR resonance lines of LiFeAs . . . 127
8.13 $^{7}$Li quadrupole frequency of LiFeAs . . . 127
8.14 $^{7}$Li NMR Knight shift of LiFeAs . . . 128
8.15 $^{7}$Li NMR $(T_1T)^{-1}$ of LiFeAs . . . 129

A.1 Theoretical powder pattern in case of first order quadrupole effects . . . 136
A.2 Theoretical powder pattern for the central transition in case of second order quadrupole effects . . . 137
A.3 Experimental $^{75}$As NMR powder pattern for LiFeAs . . . 138
A.4 Stretched exponential fitting for $LaO_{0.9}F_{0.1}FeAs_{1-\delta}$ at $T = 22\,K$ . . . 144
A.5 $T_1^{-1}$ of $LaO_{0.9}F_{0.1}FeAs_{1-\delta}$ for different stretched exponential functions . . . 146
A.6 Spin diffusion . . . 148
A.7 Field dependence of $(T_1T)^{-1}_{sc}$ at $T = 8$ K . . . 149

# List of Tables

4.1 Values of the superconducting gap(s) in $LaO_{1-x}F_xFeAs$ . . . . . . . . . . . . 50

5.1 Transition temperatures $T_s$, $T_N$ and $T_c$ of $LaO_{1-x}F_xFeAs$ . . . . . . . . . . 56

6.1 Hyperfine couplings and orbital shifts for $LaO_{0.9}F_{0.1}FeAs$ . . . . . . . . . . . 72
6.2 Curie-Weiss fitting parameters for underdoped $LaO_{1-x}F_xFeAs$ . . . . . . . 87
6.3 BPP fitting parameters for underdoped $LaO_{1-x}F_xFeAs$ . . . . . . . . . . . . 89

8.1 Summary of NQR and NMR properties of all LiFeAs samples . . . . . . . 125

A.1 $T_1$ values for $LaO_{0.9}F_{0.1}FeAs_{1-\delta}$ in $H = 7.01\,\mathrm{T}$ derived from different stretched exponential fitting functions . . . . . . . . . . . . . . . . . . . . . . . 145

# 1 Introduction

One hundred years after its discovery by Heike Kamerlingh Onnes in April 1911 in Leiden [1], superconductivity is still a fascinating and mysterious topic. Nearly half a century had to pass by until its mechanism in the conventional superconductors with rather low transition temperatures ($T_c$) could be explained consistently within the theory by Bardeen, Cooper and Schrieffer (BCS), published in 1957 [2]. Nowadays, most of the unconventional superconductors such as heavy fermions, discovered in 1979 [3], and the famous high temperature cuprate superconductors, discovered in 1986 [4], are far from being understood completely.

In February 2008, Kamihara *et al.* reported the discovery of superconductivity in $LaO_{1-x}F_xFeAs$ with a transition temperature up to $T_c = 26\,K$ [5] and thus opened the stage for a new family of superconducting materials: the iron-based superconductors (also called pnictides). Immediately afterwards, the superconducting research community world-wide started an extensive quest for similar compounds and for the understanding of the mechanism of superconductivity in these materials. By replacing La with some other rare earth (RE) elements and either fluorine doping or the introduction of oxygen deficiencies, superconducting transition temperatures up to 55 K were found in these so-called "1111" systems [6–10]. Aside from the cuprates, these are the highest transition temperatures among all known superconductors. Other subfamilies such as the "122" systems (e.g. $BaFe_2As_2$ [11–13]), the "111" (e.g. LiFeAs [14–16]) and the "11" (e.g. FeSe [17]) with similar but somewhat lower superconducting transition temperatures were discovered within a relatively short period of time, leading to a second heyday of high temperature superconductivity research, similar to the days after the discovery of the cuprates in 1986 [4].

Nuclear Magnetic Resonance (NMR) has been a fundamental player in the studies of superconducting materials for many decades. This local probe technique takes advantage of the hyperfine coupling mechanism of the nuclear spins to the electrons (quasiparticles) in the normal (superconducting) state. This reveals information about the pairing state of the Cooper pairs, the symmetry of the superconducting order parameter as well as information about normal state properties. To name but a few, the observation of the Hebel-Slichter peak in the spin-lattice relaxation rate of superconducting aluminum was a very important confirmation of the BCS theory [18–20] and NMR measurements on $Sr_2RuO_4$ and the heavy fermion superconductor $UPt_3$ provided experimental evidence for unconventional spin-triplet superconductivity [21, 22]. Electronic correlations in the normal state can also be well investigated by means of NMR. In the high-temperature cuprate superconductors, for instance, NMR evidenced the presence of antiferromagnetic correlations as well as the opening of a pseudogap in the underdoped regime [23]. In the field of the recently discovered iron-based superconductors, NMR has been widely applied since the very beginning of the pnictide research. This thesis comprises some of these very first results of NMR measurements on pnictides.

In the course of this work, NMR and NQR (nuclear quadrupole resonance) measurements on two different families will be presented: $LaO_{1-x}F_xFeAs$, a member of the 1111-family, and LiFeAs. $LaO_{1-x}F_xFeAs$ was the first pnictide material for which superconductivity was reported [5]. It was the first available compound and thus studied in detail. LiFeAs, on the other hand, is very special since it is a stoichiometric superconductor.

This thesis is organized as follows: After an introduction to the experimental method of NMR in Chapter 2, a short review of NMR measurements in the superconducting state and what can be learned from them will be given in Chapter 3. Both investigated compounds and their relation compared to other pnictides, as well as a comparison between cuprates and pnictides in general, will be described in Chapter 4. The detailed experimental setup is given in Chapter 5, including some problems faced during the experiments and their solutions/improvements. The presentation and discussion of the NMR measurements on $LaO_{1-x}F_xFeAs$ is divided into two parts. Chapter 6 reports NMR measurements in the normal state of $LaO_{1-x}F_xFeAs$ and discusses the data regarding the single spin fluid character of the multi-band electronic structure, the value of the ordered moment in the magnetically ordered state, the role of spin flucuations for the occurrence of superconductivity, and the Fermi-liquid/non-Fermi-liquid regions in the phase diagram. Chapter 7 concentrates on the low-energy spin dynamics in the superconducting state of $LaO_{1-x}F_xFeAs$ and of a sample with an artificially enhanced impurity concentration: $LaO_{0.9}F_{0.1}FeAs_{1-\delta}$. The temperature dependence of the spin-lattice relaxation rate will be discussed within possible superconducting gap symmetries, by taking into account the results of other experimental methods as well as the effect of impurities. NMR and NQR measurements on three different single crystals and a polycrystalline sample of LiFeAs will be overviewed in Chapter 8. A strong variation of the results in the normal state as well as in the superconducting state indicates the proximity of LiFeAs to an instability, where subtle changes in the stoichiometry may lead to completely different ground states. Chapter 9 recapitulates the outcome of the performed measurements and gives an outlook for possible further investigations.

# 2 Basic Principles of NMR

Nuclear magnetic resonance (NMR) is the oldest nuclear method in solid state physics. It is based on the principle, that transitions between nuclear magnetic energy levels corresponding to differently oriented nuclear spins in a static magnetic field should be observable when applying a second, time-dependent magnetic field perpendicular to the static one. The second magnetic field should oscillate at the Larmor frequency of the nuclei. The first NMR measurements were performed contemporaneously by Purcell, Torrey and Pound in Cambridge and by Bloch, Hansen and Packard in Stanford. Purcell and his colleagues observed the radio frequency absorption of protons in solid paraffin at room temperature using a resonant cavity, a sweepable magnet and radio frequency power of about $10^{-11}$ W [24]. At the same time Bloch, Hansen and Packard reported in a very short paper ($\sim$ 260 words) the observation of radio frequency absorption by protons in water at room temperature using conventional radio frequency techniques [25]. Bloch, who was not limited to a fixed cavity frequency, could already measure at different frequencies and fields and confirmed that the ratio $H/\nu$ was always the same: the gyromagnetic ratio $\gamma$ of the protons. In his paper, which was published on Christmas Eve of 1945, Purcell already proposed various applications of the newly established effect, for instance precise measurements of gyromagnetic ratios, investigations of the spin-lattice coupling as well as standardizations of magnetic fields. All these applications (and many more) were realized sooner or later. This was the twofold birth of NMR, which already included the basic principle of NMR based on the isotope-specific gyromagnetic ratios as well as first indications of the broad applicability of NMR in condensed matter and beyond. For completeness and to honor the nowadays unfortunately unpopular practice to publish also negative results, it should be noted that an earlier attempt to detect nuclear magnetic resonance four years before the discoveries of Bloch and Purcell was unsuccessful [26]. Over the years, NMR became a powerful method in condensed matter as well as in in chemistry, biology and medicine, where it is widely used for structural analysis and non-destructive diagnostic imaging.

The most distinctive characteristic of NMR and its special advantage compared to many other methods in solid state physics is its local probe character. Different magnetic and electronic environments of the nuclei will lead to different NMR responses on both the static and/or dynamic side. In contrast to macroscopic susceptibility or magnetization measurements, with which one can only probe the whole magnetic response of a system (the bulk susceptibility including impurity contributions), NMR can distinguish intrinsic contributions to the susceptibility from different magnetic ions in the material as well as impurity contributions. An advantage of NMR over $\mu$SR (muon spin rotation), which is also a local probe technique, is the fact, that by doing NMR one knows exactly at which nuclear site one is probing the system, while in $\mu$SR one always has to calculate (or to guess) at which (possibly interstitial) site the muon is sitting. By doing NMR on different elements, one can deduce their different couplings to the electronic environment. A partial contribution to the intrinsic susceptibility might be small, but due to its high sensitivity to small magnetic moments down to $\mu = 10^{-2}\mu_B$ and its extremely high relative frequency resolution better than $10^{-7}$ (in high resolution NMR of liquids [27]), NMR can still detect this contribution and separate it from other contributions. A famous example is the measurement of the diminutive nuclear paramagnetism in ferromagnetic iron [28]. Another example, taken from our studies at the IFW Dresden, is high temperature superconducting $NdBa_2Cu_3O_{6+y}$, where one can distinguish between the Cu-sites in the chains and the Cu-sites in the $CuO_2$ planes and study their properties selectively by means of NMR and nuclear quadrupole resonance (NQR) [29]. An example of the local probing of the charge environment are underdoped $LaO_{1-x}F_xFeAs$ and $SmO_{1-x}F_xFeAs$, where by means of NQR two different types of charge environments of the $^{75}As$ nuclei, coexisting at the nanoscale and thus pointing to a local electronic order in these systems, have been found [30]. As a fourth and last example NMR measurements on the high-temperature superconductor $YBa_2Cu_3O_{6.63}$ will be given [31]. Local susceptibilities of different nuclei have been measured in this compound and it has been found that all components show the same temperature behavior and can even be scaled with the temperature dependence of the macroscopic susceptibility. These results showed that all nuclei probe the same single spin component of the system, despite the existence of Cu $3d$ and O $2p$ states in this material. On the other hand, the spin-lattice relaxation rates of $^{63}Cu$ and $^{17}O$ in the same system differ from one another, because depending on their form factors, different nuclei probe different regions in $q$-space: $^{63}Cu$ can probe antiferromagnetic correlations, while $^{17}O$ is "blind" for spin fluctuations at $Q = (\pi, \pi)$. So by choosing the right nucleus, one can not only probe different spatial regions in the crystal structure, but also different regions in reciprocal space.

These four examples reveal the rich variety of possible applications of the NMR gold mine in solid state physics. Most of these applications (probing local susceptibilities, probing electronic environments, probing different $q$-spaces) will play a role in the course of this thesis.

The following description of the basic principles of NMR and NQR is based on the textbooks by C.P. Slichter [28], A. Abragam [32], E. Fukushima and S.B.W. Roeder [33], and G. Schatz and A. Weidinger [27], as well as on selected Ph.D. theses [34–38] and original publications.

## 2.1 Isolated Nuclear Spins in a Magnetic Field

### 2.1.1 A Single Nuclear Spin

Most of the elements of the periodic table (namely all besides the ones with an even number of protons and an even number of neutrons in their nucleus) possess an isotope-specific nuclear spin $\vec{I}$. The nuclear spin causes a nuclear magnetic moment $\vec{\mu}$ proportional to this spin:

$$\vec{\mu} = \frac{g\mu_N}{\hbar}\vec{I} = \gamma\vec{I}. \tag{2.1}$$

Here, $g$ denotes an effective $g$-factor [27], $\mu_N = 5.05 \cdot 10^{-27}$ A.m$^2$ is the nuclear magneton, and $\hbar$ is the Planck constant. The gyromagnetic ratio $\gamma$ combines all the prefactors of

Eq. (2.1) to an isotope-specific constant, ranging from $\gamma/2\pi = 42.5749\,\mathrm{MHz/T}$ for $^1$H to $\gamma/2\pi = 0.7291\,\mathrm{MHz/T}$ for $^{197}$Au.[1]

The nuclear magnetic moment couples to an external magnetic field[2] $\vec{H}_0$ via the Zeeman interaction, given by the following Hamiltonian:

$$\mathcal{H}_Z = -\vec{\mu} \cdot \vec{H}_0 \,. \tag{2.2}$$

In quantum mechanics, $\vec{I}$ has to replaced by the operator $\hat{I}$ with the eigenvalues $\hbar m$. Assuming that the external field is aligned along the $z$-direction: $\vec{H}_0 = H_0 \vec{e}_z$, one arrives at:

$$\mathcal{H}_Z = -\gamma H_0 \hat{I}_z \,, \tag{2.3}$$

and finally gets the corresponding $(2I+1)$ equidistantly split energy levels:

$$E_m = -\gamma H_0 \langle I, m | \hat{I}_z | I, m \rangle = -\gamma \hbar m H_0 \tag{2.4}$$

with

$$m = -I, -I+1, ..., I-1, I \,. \tag{2.5}$$

Since the matrix elements between states $|\hat{I}_z, m\rangle$ and $|\hat{I}_z, m'\rangle$ of the operator $\hat{I}_z$ vanish unless $m' = m \pm 1$, transitions between these levels are allowed only for $\Delta m = \pm 1$. To undergo such a transition, the system has to absorb the energy $\Delta E = \hbar \omega_L$, where

$$\omega_L = \gamma H_0 \tag{2.6}$$

is the so-called Larmor frequency, with which the nuclear spin is precessing around the static magnetic field $H_0$.

### 2.1.2 Many Nuclear Spins

The previous derivation holds basically for a single nuclear spin. However, the observation of a radio frequency absorption by a single nuclear spin is unrealistic. To do NMR, we need around $10^{16}$ nuclear spins for reasonable signal-to-noise ratios[3]. In thermodynamic equilibrium, the nuclear magnetic energy levels $E_m$ given by Eq. (2.4) are occupied obeying the Boltzmann distribution

$$P(m) = \exp\left(\frac{-E_m}{k_B T}\right) \,. \tag{2.7}$$

[1] In practice, $\gamma$ is mostly expressed in terms of $\gamma/2\pi$, since this value is used to calculate the radio frequency $\nu_L = \omega_L/2\pi = \gamma H_0/2\pi$ with which the system has to be irradiated such that it absorbs the energy $E = \hbar\omega_L = h\nu_L$. For theoretical considerations however, one mainly uses the angular frequency $\omega_L = \gamma H_0$ with which the spins precess around the direction of the external field. For $^1$H $\gamma$ amounts to $26.7522128 \times 10^7\,\mathrm{rad/(s.T)}$ [39].

[2] The magnetic field strength $H$ has the units A/m. A correct description would use the magnetic flux density $B = \mu_0 H$ in units of T for the description of the magnetic field. However, in order to keep consistency with textbooks [28, 32, 33] and most of the literature, $H$ will be used for the magnetic field in this thesis. See also discussion "$B$ vs $H$" in [33].

[3] Nowadays, specially assembled high resolution NMR can reach much lower limits of detection. For instance, magnetic resonance force microscopy (MRFM) is able to detect down to $10^6$ nuclear spins [40]. This number is already at a stage were statistical spin fluctutations rather than the Boltzmann distribution are dominant.

The resulting unequal occupation of the energy levels leads to a nuclear spin polarization $\langle \hat{I}_z \rangle$ and hence to a nuclear magnetization $M_z$:

$$M_z = N\gamma\langle \hat{I}_z \rangle = N\gamma \frac{\sum_{m=-I}^{m=+I} \hbar m \exp\left(-E_m/k_B T\right)}{\sum_{m=-I}^{m=+I} \exp\left(-E_m/k_B T\right)} \approx \frac{N\gamma^2\hbar^2 I(I+1)}{3k_B T} H_0 = \chi_n H_0 \,, \tag{2.8}$$

where $N$ denotes the density of nuclear spins and $\chi_n$ the static nuclear susceptibility. The penultimate part of Eq. (3.21) was obtained by linearly expanding the Boltzmann exponential, since $E_m/k_B T$ is always small. The nuclear spin polarization $\langle \hat{I}_z \rangle$ is rather weak[4]. However, this small difference in the occupation of the nuclear magnetic energy levels is sufficient to enable NMR measurements. NMR cannot observe single nuclear spins, but measures the nuclear magnetization $M_z$, which is a collective property of the nuclear spin system. In the future we will therefore not use the picture of a single spin precessing with the Larmor frequency $\omega_L$ around the direction of the static magnetic field, but the picture of a precessing nuclear magnetization $M_z$. According to Eq. (3.21), $M_z$ increases with an increasing number of nuclear spins, $N$, with increasing the external magnetic field $H_0$ and by lowering the temperature. These dependences already contain very important information for the performance of the measurements. Since $\gamma$ contains the nuclear magneton, which is already much smaller than the Bohr magneton $\mu_B$ ($\mu_N \sim 0.0005\mu_B$), and enters squared in Eq. (3.21), the static nuclear susceptibility is much smaller than any electronic one (about a factor of $10^{-6}$ to $10^{-8}$). The big difference between the nuclear and the electronic susceptibility is fundamental for the application of NMR in condensed matter. As we will see in the next Section, nuclear spins and electronic spins interact with each other via hyperfine couplings. It is therefore possible to study the properties of the electronic spin system by observing the response of the nuclear spin system, whose influence on the electronic spin system is negligibly small.

## 2.2 Making Life Interesting: Interactions

Everything told in the previous Section would be rather bland without interactions, since the resonance frequency $\omega$ of an isotope in a given static magnetic field $H_0$ would correspond to the Larmor frequency $\omega_L$ in this field deduced from the well-known gyromagnetic ratios $\gamma$ and Eq. (2.6). Interactions among the nuclear spins and with their electronic environment will however shift and broaden the nuclear magnetic energy levels $E_m$ and lead to relaxation effects[5]. The Hamiltonian expands to:

$$\mathcal{H} = \mathcal{H}_z + \mathcal{H}_{n-n} + \mathcal{H}_{n-e} + \mathcal{H}_q \,, \tag{2.9}$$

where $\mathcal{H}_z$ denotes the already introduced Zeeman interaction between a nuclear moment and a static magnetic field, $\mathcal{H}_{n-n}$ is the interaction between nuclear spins themselves,

[4] At $T = 300\,\mathrm{K}$, for $I = 1$ and $\mu_0 H_0 = 1\,\mathrm{T}$ $\langle \hat{I}_z \rangle$ amounts only to $10^{-6}$.

[5] The following deductions assume a single, homogeneous environment of all nuclei. A distribution of different environments would also result in line shifts, line broadening and relaxation effects.

$\mathcal{H}_{n-e}$ is the interaction between the nuclear spins and the spins and orbital momenta of the electrons and $\mathcal{H}_q$ is the electric quadrupole interaction. They will all be discussed in detail in the following.

### 2.2.1 Interactions Between Nuclear Spins - Line Shape I

The internuclear coupling $\mathcal{H}_{n-n}$ between two nuclear spins $\vec{I}_i$ and $\vec{I}_j$ can be divided into two parts, a direct dipolar coupling between nuclear spins and an indirect coupling:

$$\mathcal{H}_{n-n} = \sum_{i,j} \vec{I}_i \mathbf{a}(\vec{r}_{ij}) \vec{I}_j = \mathcal{H}_{n-n}^{dir} + \mathcal{H}_{n-n}^{indir} . \tag{2.10}$$

**The general direct dipolar Hamiltonian for $N$ nuclear spins** can be written as

$$\mathcal{H}_{n-n}^{dir} = \sum_{i<j}^{N} \frac{\hbar^2\gamma^2}{r_{ij}^3} \left[ \vec{I}_i \cdot \vec{I}_j - 3\frac{(\vec{I}_i \cdot \vec{r}_{ij})(\vec{I}_j \cdot \vec{r}_{ij})}{r_{ij}^2} \right] , \tag{2.11}$$

where $\vec{r}_{ij}$ is a vector from $\vec{I}_i$ to $\vec{I}_j$. This coupling between nuclear spins gives rise to a homogeneous broadening of the resonance line. In a rigid lattice, the dipolar linewidth is of the order of the local field produced at the site of a nuclear spin $\vec{I}_i$ by its neighbouring nuclear spins $\vec{I}_j$. The dipolar linewidth is usually of the order of a few Gauss [32] and independent of the applied magnetic field [41].

**The indirect coupling between nuclear spins** $\mathcal{H}_{n-n}^{indir}$ can be written as

$$\mathcal{H}_{n-n}^{indir} = \vec{I}_i \mathbf{a}_{ij} \vec{I}_j , \tag{2.12}$$

where the exact form of the coupling tensor $\mathbf{a}_{ij}$ depends on the detailed coupling mechanism. The indirect coupling between nuclei is mediated via electrons in the following way: a nuclear spin interacts with a surrounding electron via hyperfine interactions (see Section 2.2.2). Hyperfine interactions between this electron and a second nuclear spin leads to an indirect coupling between the two nuclear spins. Depending on the specific situation, this indirect coupling can lead to a narrowing or a broadening of the resonance line (see [28] for details). In liquids the bonding electrons of the molecules act as centers of communication between the nuclear spins, leading to the so-called $J$-coupling in high resolution NMR on liquids [42, 43]. In metallic solids the indirect coupling between nuclear spins is mediated via the coupling of the nuclear spins to the conduction electrons (so-called RKKY[6] interaction) [44, 45]. In this case the prefactor $a_{ij}$ of the *scalar* part of Hamiltonian (2.12) becomes [46, 47]:

$$a_{\alpha\alpha}(\vec{r}_{ij}) = \frac{1}{(\hbar\gamma_e)^2} \sum_{\vec{q}} \chi'_\alpha(\vec{q}) |A_{\alpha\alpha}(\vec{q})|^2 \exp(-i\vec{q}\vec{r}_{ij}) , \tag{2.13}$$

with $\alpha = x, y, z$. It is proportional to the real part of the static, $q$-dependent susceptibility $\chi'_\alpha(\vec{q})$ of the electronic system, which mediates the coupling between the nuclear spins via the hyperfine coupling $A_{\alpha\alpha}(\vec{q})$.

[6] RKKY = Ruderman, Kittel, Kasuya and Yosida

The analysis of the spin-spin relaxation rate $(1/T_{2G})^2 \propto |a(\vec{r}_{ij})|^2$ [46–49] in conductors is therefore a measure of the static, $q$-dependent susceptibility $\chi'(\vec{q})$ and thus complementary to measurements of the spin-lattice relaxation rate $T_1^{-1}$, which itself yields information about the imaginary part of the dynamic, $q$-dependent susceptibility $\chi''(\vec{q},\omega)$ (see Section 2.3.2.1).

This relation can be mainly used in the field of cuprate research, where the following conditions are met [46, 48]: First, due to the highly anisotropic hyperfine couplings in cuprate superconductors, one can omit the $x$ and $y$ components in Eq. (2.13) and obtains $(1/T_{2G})^2 \propto |a_{zz}(\vec{r}_{ij})|^2 \propto (\chi'_z(\vec{q}))^2$ (for $H_0 \parallel z$) [46–48]. Furthermore, the direct dipolar nuclear spin-spin interaction, which also contributes to $1/T_{2G}$, but does not involve interactions with the electronic spin system (see Section 2.3.3), decreases rapidly with $1/r_{ij}^3$, [see Eq. (2.11)]. In cuprates, this is negligibly small compared to the indirect nuclear spin-spin coupling, which couples all nuclear spins within the antiferromagnetic correlation length $\xi$ together [46, 48]. This assumption is based on the model of Millis, Monien and Pines (MMP model) [50], who described the $CuO_2$ planes in cuprates as an antiferromagnetic Fermi liquid, where spin fluctuations are strongly peaked at $(\pi,\pi)$. About 30 nuclear spins are then coupled together within the antiferromagnetic correlation length $\xi$ [48]. Thus $1/T_{2G}$ is dominated by the indirect nuclear spin-spin coupling and therefore a direct measure of $\chi'_z(\vec{q})$ and $\xi$. The good agreement between the theoretical description of $1/T_{2G}$ within the MMP model and the experimental observations have been a strong evidence for the MMP model and the single spin fluid description in cuprates [48], similar to the Knight shift scaling [31].

## 2.2.2 Magnetic Hyperfine Interactions

The interaction $\mathcal{H}_{n-e}$ includes the interaction of the nuclear spin with the spins and the orbital moments of the electrons via hyperfine coupling mechanisms. Together with the first term of Eq. (2.9), the Zeeman term, one obtains:

$$\mathcal{H}_Z + \mathcal{H}_{n-e} = -\gamma\hbar\vec{I}(1+\mathbf{K})\vec{H}_0\,, \tag{2.14}$$

with $\mathbf{K}$ being the Knight shift tensor. For a measurement along $H_0 \parallel z$, the resonance frequency is shifted from the expected Larmor frequency $\omega_L$ of a free nucleus to the observed resonance frequency $\omega_{obs} = (1+K_z)\omega_L$ by the Knight shift $K_z$.

Historically, the Knight shift $K$ describes hyperfine interactions between the nuclear spins and the conduction electrons. It was described in 1949 by W. D. Knight. He observed that the NMR resonance lines in metals experienced a much larger shift in the same magnetic field than their non-magnetic salts [51]. This is due to the fact that a nucleus in a metal couples to the overall paramagnetic Pauli susceptibility of the conduction electrons in a magnetic field, which is given by:

$$\chi_P = \frac{M}{H_0} = \frac{3N\mu_B^2}{2k_BT_f}\,, \tag{2.15}$$

where $M$ is the electronic magnetization which arises from the difference of electronic spins being aligned parallel or antiparallel to the applied magnetic field $H_0$, $T_f$ is the Fermi temperature and $N$ the electron density. The Pauli susceptibility is thus temperature-independent, as long as the electronic density of states is temperature-independent.

Nowadays, the Knight shift $K$ (then more precisely called NMR shift) is used with a broader meaning, covering all possible couplings between nuclear spins and electronic spins, including electrons from inner closed shells and from unfilled shells, whether itinerant (conduction electrons) or localized. The different contributions are divided into the diamagnetic NMR shift $K_{dia}$, the orbital shift $K_{orb}$ and the spin shift $K_s$:

$$K = K_{dia} + K_{orb} + K_s \,. \tag{2.16}$$

They are described in detail in the following.

## The Hyperfine Hamiltonian

The hyperfine Hamiltonian $\mathcal{H}_{hf}$ describes the interaction between a nuclear spin $\vec{I}$ and the spin $\vec{S}$ and the momentum $\vec{L}$ of an electron being at a distance $\vec{r}$ from the nucleus:

$$\mathcal{H}_{hf} = \gamma_n \gamma_e \hbar^2 \vec{I} \cdot \left[ \left( 3\frac{(\vec{S}\cdot\vec{r})\cdot\vec{r}}{r^5} - \frac{\vec{S}}{r^3} \right) + \frac{8\pi}{3}\vec{S}\delta(\vec{r}) + \frac{\vec{L}}{r^3} \right] . \tag{2.17}$$

It comprises the couplings causing the orbital shift $K_{orb}$ and $K_s$.

### The Orbital Shift

The last term of Eq. (2.17) represents the **orbital interaction** between the nuclear spin $\vec{I}$ and the angular momentum $\vec{L}$ of the electron. In contrast to the diamagnetic shift, which results from current loops in inner orbitals, where the angular momentum is quenched, the orbital shift $K_{orb}$ is due to current loops from electrons which possess a static angular momentum $\vec{L}$. For transition metal ions in crystal fields the orbital moment of the conduction electrons is also quenched. An externally applied magnetic field can however partially reconstruct an orbital moment by mixing the ground state with low lying excited states [36]. This is the so-called Van Vleck paramagnetism which leads to an orbital shift of the form [49]:

$$K_{orb} = 2\langle r^{-3} \rangle \chi_{VV} \,, \tag{2.18}$$

where $\chi_{VV}$ is the (Van Vleck) paramagnetic susceptibility. This susceptibility is usually temperature-independent. $K_{orb}$ is thus temperature-independent, as well.

### The Spin Shift

The first term (inside the parentheses) describes the **dipolar interaction** between the nuclear spin and the electronic spin leading to a dipolar shift $K_{dip}$. A (point)-dipolar approximation can however only be made if the distance between the nuclear spin and the electronic spin is large enough. This is the case for $p$-, $d$- and $f$-orbitals, but not for $s$-electrons, which exhibit a finite spin density at the nuclear site. The direct coupling between the nuclear spin and the spin of unpaired $s$-electrons is described by the remaining term in Eq. (2.17). It is called the **Fermi contact contribution**. The shift resulting from this interaction, $K_{contact}$, is proportional to the probability of finding the $s$-electron at the nuclear site, expressed by the squared modulus of the electronic wave function at $\vec{r} = 0$: $|\psi(\vec{r} = 0)|^2$. Spins from unfilled outer electronic shells will polarize even closed $s$-shells, leading to the so-called **core polarization** term $K_{core}$, which acts in the same way as the Fermi contact contribution does, but with opposite sign. $K_{core}$ is negative, while $K_{contact}$ is positive. Both terms are isotropic, while the dipolar interaction and the resulting $K_{dip}$ depend on the relative orientation of the crystal to the applied magnetic field and are therefore anisotropic. $K_{contact}$ and $K_{core}$ are subsumed under the isotropic part of the NMR shift, $K_{iso}$, since they are usually not distinguishable one from another. All three contributions $K_{dip}$, $K_{contact}$, and $K_{core}$ are only caused by the interactions between the spin of the nucleus $\vec{I}$ and the spin of the electron $\vec{S}$. No angular momentum $\vec{L}$ is involved in these three shifts. They are therefore combined to form the spin part of the NMR shift:

$$K_s = K_{dip} + K_{contact} + K_{core} \, . \tag{2.19}$$

The spin parts of the Hamiltonian (2.17), combining interactions between a nuclear spin at site $i$, $\vec{I}^i$, and an electron spin at site $j$, $\vec{S}^j$ (and thus referring to $K_s$), can be expressed as:

$$\mathcal{H}_{spin} = -\vec{I}^i \mathbf{A}^{ij} \vec{S}^j \, , \tag{2.20}$$

where $\mathbf{A}^{ij}$ is the hyperfine coupling tensor. To be more general, $\vec{S}$ can be replaced by its expectation value $\langle \vec{S} \rangle = \chi_s H_0$. It follows, that the spin part of the NMR shift $K_s$ is proportional to the static spin susceptibility $\chi_s$:

$$K_s = \frac{A^{ij}}{g_e \mu_B \gamma \hbar} \chi_s \, , \tag{2.21}$$

where $g_e = \gamma_e \hbar / \mu_B$ is the Landé factor of the electron. More precisely, the spin shift measures the static spin susceptibility at $\vec{q} = 0$: $K_s \propto A_{hf}(\vec{q} = 0)\chi(\vec{q} = 0)$. The static spin susceptibility $\chi_s$ includes the Pauli spin susceptibility of the conduction electrons $\chi_P$ [see Eq. (2.15)] as well as the paramagnetic susceptibility $\chi_{para}$ of localized electronic moments, which reads:

$$\chi_{para} = \frac{N' \mu_B^2 p^2}{3 k_B T} = \frac{C}{T} \, , \tag{2.22}$$

where $p = \sqrt{g^2 S(S+1)}$ is the effective number of Bohr magnetons and $C$ the introduced Curie constant. In contrast to the temperature-independent Pauli susceptibility $\chi_{para}$ is inversely proportional to the temperature. It can be easily larger than the Pauli susceptibility $\chi_P$, since usually $T_f/T \approx 10^2$.

Equation (2.21) reveals one of the great conveniences of NMR: by measuring the shift of the NMR resonance line, one can obtain information about the *local*, intrinsic static spin susceptibility, which is not affected by extrinsic impurity contributions. Furthermore, this information can be extracted without perturbing the electronic system while performing the measurements, since the Larmor frequency is basically zero compared to the frequencies of the electronic spin system [38]. The hyperfine coupling $A^{ij}$ depends on the different possible coupling paths from the nucleus to on-site orbitals and orbitals on surrounding atoms via transferred hyperfine interactions. If the macroscopic susceptibility $\chi_{macro}$ is not covered by extrinsic impurity effects, a scaling of $K$ versus $\chi_{macro}$ can reveal information about the hyperfine coupling at $\vec{q} = 0$ (see Section 6.1).

#### The Diamagnetic Shift

The diamagnetic shift $K_{dia}$ stems from the diamagnetism of the closed inner electronic shells (ion cores) and the Landau diamagnetism of the conduction electrons. It is temperature-independent. Diamagnetic shifts are rather small (of the order of $\Delta\omega/\omega \approx 10^{-5}$). They are only observable in high resolution NMR of liquids, where resonance lines are narrow and the resolution is up to $\Delta\omega/\omega \approx 10^{-7}$ [27]. In solid state NMR of magnetic and conducting materials other contributions to the NMR shift are usually much larger and thus the diamagnetic shift can be neglected.

### 2.2.3 Electric Quadrupole Interactions

The last term of Eq. (2.9), $\mathcal{H}_Q$ describes the electric quadrupole interaction between the quadrupole moment of the nucleus, $Q$, and the electric field gradient (EFG) $V_{\alpha\beta}$:

$$\mathcal{H}_Q = \frac{1}{6}\sum_{\alpha\beta} V^i_{\alpha\beta} Q^i_{\alpha\beta} \,. \tag{2.23}$$

**The EFG, $V_{\alpha\beta}$** is caused by the surrounding charge distribution of the nucleus. It is given by $V_{\alpha\beta} = \left.\frac{\partial^2 V}{\partial x_\alpha \partial x_\beta}\right|_{r=0}$, where $V$ is the electrostatic potential and $x_\alpha, x_\beta = x, y, z$. In its principal axes system $(X, Y, Z)$ the tensor $V_{\alpha\beta}$ can be described by its diagonal elements $V_{XX}$,$V_{YY}$ and $V_{ZZ}$. Since the potential $V$ has to fulfill Laplace's equation: $\nabla^2 V = 0$, the EFG is traceless:

$$V_{XX} + V_{YY} + V_{ZZ} = 0 \,. \tag{2.24}$$

It can therefore be described by using only two parameters. It is common to use the convention $|V_{ZZ}| \geqslant |V_{YY}| \geqslant |V_{XX}|$ and then take $V_{ZZ} \equiv eq$ and the asymmetry parameter:

$$\eta = \frac{V_{XX} - V_{YY}}{V_{ZZ}} \qquad 0 \leq \eta \leq 1 \,, \tag{2.25}$$

for the adequate description of the EFG. Eq. (2.23) then becomes:

$$\mathcal{H}_Q = \frac{1}{6}\sum_{\alpha} V_{\alpha\alpha} Q_{\alpha\alpha} \,. \tag{2.26}$$

In a cubic symmetry as well as for a spherically-symmetrical charge distribution ($s$-electrons for instance), $V_{XX} = V_{YY} = V_{ZZ}$. From Eq. (2.24) it follows that in this case $V_{XX} = V_{YY} = V_{ZZ} = 0$. The quadrupole interaction vanishes in these cases.

**The quadrupole moment of the nucleus, $Q_{\alpha\alpha}$**, describes deviations of the nuclear charge distribution from a spherically symmetric distribution. Classically it is defined as:

$$Q_{\alpha\alpha} = \int \rho(\vec{r})(3x_\alpha^2 - r^2)d^3r \,, \tag{2.27}$$

where $\rho(\vec{r})$ is the charge density at a distance $\vec{r}$ away from the origin of the nucleus. In a quantum mechanical description[7], it can be shown that $Q$ vanishes for all nuclei with $I = 0$ and $I = \frac{1}{2}$. Hence the quadrupole interaction has to be considered only for nuclei with $I \geqslant 1$ in a non-cubic environment.

**The Hamiltonian of the quadrupole interaction, $\mathcal{H}_Q$,** can be written as:

$$\mathcal{H}_Q = \frac{h\nu_q}{2}\left(I_Z^2 - \frac{I(I+1)}{3} + \frac{\eta}{6}(I_+^2 + I_-^2)\right), \tag{2.28}$$

where $Z$ refers to the $Z$-axis in the principle axis system of the EFG (which might differ from the $z$-axis in a reference system, where the magnetic field $H_0$ and thus the nuclear spin polarization are aligned along $z$). $I_+$ and $I_-$ are raising and lowering operators and $\nu_q$ is the quadrupole frequency defined as:

$$\nu_q = \frac{3eQV_{ZZ}}{2I(2I-1)h} = \frac{3e^2qQ}{2I(2I-1)h}. \tag{2.29}$$

In a strong magnetic field $H_0$, where $\nu_q \ll \nu_L = \omega_L/2\pi$, the quadrupole Hamiltonian $\mathcal{H}_Q$ can be treated as a pertubation of the Zeeman term $\mathcal{H}_Z$.

In first order the change of energy of the $m$th energy level is given by:

$$E_m^{(1)} = \frac{eQV_{zz}}{4I(2I-1)}[3m^2 - I(I+1)], \tag{2.30}$$

where $V_{zz}$ is the component of the EFG parallel to the direction of the applied magnetic field. Expressing $V_{zz}$ in the principal axis system of the EFG by using the Euler angles $\theta$ and $\phi$ which describe the orientation of the principle axis system of the EFG with respect to the laboratory axis system of the external magnetic field[8]:

$$V_{zz} = V_{XX}\sin^2\theta\cos\phi + V_{YY}\sin^2\theta\sin\phi, +V_{ZZ}\cos^2\theta \tag{2.31}$$

and inserting the definitions of the asymmetry parameter $\eta$ and the quadrupole frequency $\nu_q$ given in Equations (2.25) and (2.29), one arrives at [32]:

$$E_m^{(1)}(\theta,\phi,\eta) = \frac{1}{4}h\nu_q\left(3\cos^2\theta - 1 + \eta\sin^2\theta\cos 2\phi\right)\left[m^2 - \frac{1}{3}I(I+1)\right]. \tag{2.32}$$

These energy changes result in a splitting of the NMR resonance line into $2I$ resonance lines, where adjacent lines are separated by a frequency proportional to $\nu_q$, whose absolute value depends on the orientation of the principle axes of the EFG relative to the direction of the applied magnetic field. This first-order quadrupole shift is given by the difference $E_{m-1}/h - E_m/h$ and reads [32]:

$$\Delta^{(1)}\nu_m(\theta,\phi,\eta) = -\frac{\nu_q}{2}\left(3\cos^2\theta - 1 + \eta\sin^2\theta\cos 2\phi\right)(m - 1/2). \tag{2.33}$$

[7] One arrives at the quantum mechanical description by simply replacing $\rho$ with its quantum mechanical operator $\hat{\rho}(\vec{r}) = \sum_k q_k\rho(\vec{r}-\vec{r}_k)$ where the sum runs over the nuclear particles 1,2,...$k$, ...N with charge $q_k$ and position $\vec{r}_k$.

[8] $\theta$ is the angle between the principal axis of the EFG, $Z$, and the direction $z$ of the applied magnetic field, and $\phi$ denotes a rotation around $z$ in the $xy$-plane.

Note that the resonance frequency of the central transition $(m = -\frac{1}{2}) \longleftrightarrow (m = +\frac{1}{2})$ remains unaffected in first order, since according to Eq. (2.32) both levels $(m = \pm\frac{1}{2})$ are shifted by the same amount of energy (see also Fig. 2.2). The intensity of each of the $2I$ resonance lines depends on $m$.

The change of energy of the $m$th energy level in second order, $E_m^{(2)}$, can be found for instance in [32]. For $\eta = 0$, $E_m^{(2)}$ is odd in $m$, which means that the distance between the satellites does not change when considering second-order effects. It is still the same as in first order [32, 49]. Only the resonance frequency of the central transition shifts.[9] Its second-order quadrupole shift is [32, 54–56]:

$$\Delta^{(2)}\nu_m(\theta,\phi,\eta) = -\frac{\nu_q^2}{\nu_L}\frac{1}{6}\left[I(I+1) - \frac{3}{4}\right]\left[A(\phi,\eta)\cos^4\theta + B(\phi,\eta)\cos^2\theta + C(\phi,\eta)\right]. \tag{2.34}$$

The prefactors $A, B$ and $C$ depend on the angle $\phi$ and the asymmetry parameter $\eta$ of the EFG [54–56] (see Appendix A.1).

Higher orders are usually not taken into account. In general, the central line is only affected by even-value order terms of the perturbation theory, while the quadrupole shifts of the satellites only depend on odd orders of the pertubation [57]. Thus, the next relevant order for the satellites would be the third one, leading to a shift proportional to $\nu_q^3$ [57]:

$$\Delta^{(3)}\nu_m = \frac{\nu_q^3}{3\nu_L^2}(2m-1)P_3(m), \tag{2.35}$$

where the prefactor $P_3(m)$ contains the dependency on the components of the EFG and therewith the dependency on the orientation [57]. The next interesting order for the central line would be the fourth one, leading to a shift of the order of $\mathcal{O}(\nu_Q^4/\nu_L^3)$, which is so small compared to the second-order shift [58, 59], that it is commonly neglected and no explicit expression for it could be found in literature.

Please note that all preceding statements referred to the case of NMR on single crystals, where a certain orientation of the crystal axes and thus of the EFG versus the magnetic field can be chosen. For the more complicated case of NMR on polycrystalline powder samples please refer to Appendix A.1.

### Nuclear Quadrupole Resonance

Even in the absence of an external magnetic field, the quadrupole interaction $\mathcal{H}_Q$ leads to a splitting of the nuclear magnetic energy levels, where energy levels with the same $|m|$ are shifted equally. Resonant transitions between these twofold degenerate energy levels can be induced by applying an alternating magnetic field $H_1$ (a radio frequency pulse with the corresponding frequency) in the direction perpendicular to the principal axis of the EFG. This is widely used in the field of *Nuclear Quadrupole Resonance* (NQR), which

[9] This holds only true for NMR spectra on single crystals. In the case of powder spectra the situation is more complicated. (Asymmetric) second-order quadrupole effects ($\eta \neq 0$) have to be considered also for the satellites of the powder spectrum [52]. Furthermore, additional satellites due to $\theta \neq 90°$ contributions might appear [53].

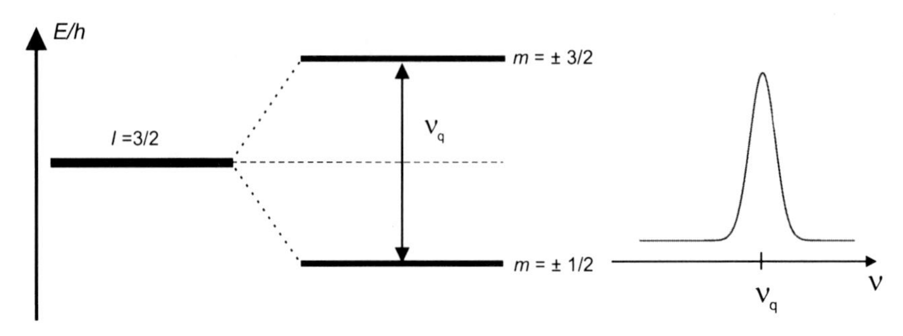


**Figure 2.1:** Level splitting and the corresponding NQR spectrum due to quadrupole effects without an external magnetic field for $I = 3/2$ and $\eta = 0$. The fourfold degenerate energy levels (left side) split in the presence of a non-vanishing EFG into twofold degenerate energy levels, separated by the quadrupole frequency $\nu_q$ (center), resulting in one single NQR resonance line observable at $\nu_q$ (right side).

was first performed in 1955 [60] and allows to study the charge environment of a nucleus without having to consider additional magnetic interactions.[10] For half-integer nuclear spins $I > 3/2$ and $\eta = 0$, $I - 1/2$ resonance lines with frequencies proportional to $\nu_q$, $2\nu_q$, $3\nu_q$ ... will be observed.

For example, the NQR frequency of a nuclear spin $I = 3/2$ in zero magnetic field is given by:

$$\nu_{\mathrm{NQR}} = \nu_q \sqrt{1 + \frac{\eta^2}{3}} = \frac{3eQV_{ZZ}}{2I(2I-1)h} \sqrt{1 + \frac{\eta^2}{3}} \,. \tag{2.36}$$

In a uniaxial symmetry ($\eta = 0$) it equals the quadrupole frequency $\nu_q$ (see Fig. 2.1).

### Example of $I = 3/2$

The example of a nuclear spin $I = 3/2$ in an uniaxial symmetry ($\eta = 0$) and in a large applied magnetic field will be considered as a special case in the following, because it will play a role in the discussion of the data presented in this thesis (see Fig. 2.2).

The shift of the nuclear magnetic energy levels $E_m$ due to first-order quadrupole effects in a uniaxial symmetry can be calculated to:

$$E_m^{(1)}(\theta, \phi, \eta) = \frac{1}{4} h\nu_q (3\cos^2\theta - 1) \left[ m^2 - \frac{1}{3} I(I+1) \right] \,. \tag{2.37}$$

This leads to a first-order quadrupole shift (distance between the unshifted central line and the satellites) of the form of:

$$\Delta^{(1)} \nu_m(\theta) = -\frac{\nu_q}{2} (3\cos^2\theta - 1)(m - 1/2) \,. \tag{2.38}$$

[10] This is only true for static measurements. The NQR spin-lattice relaxation can still be of magnetic origin.

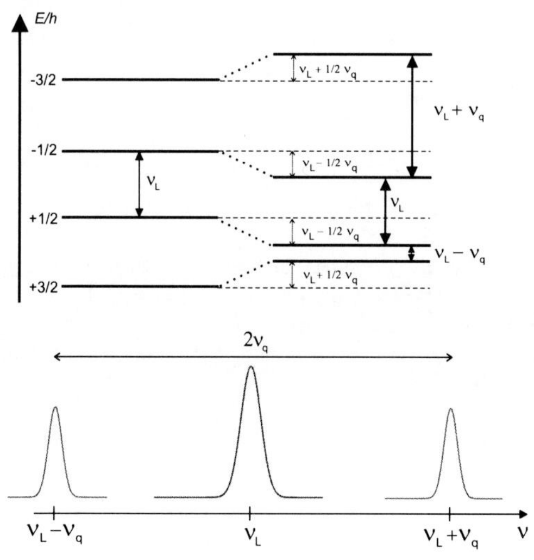


(a) Level splitting and corresponding NMR spectrum for $\theta = 0°$.

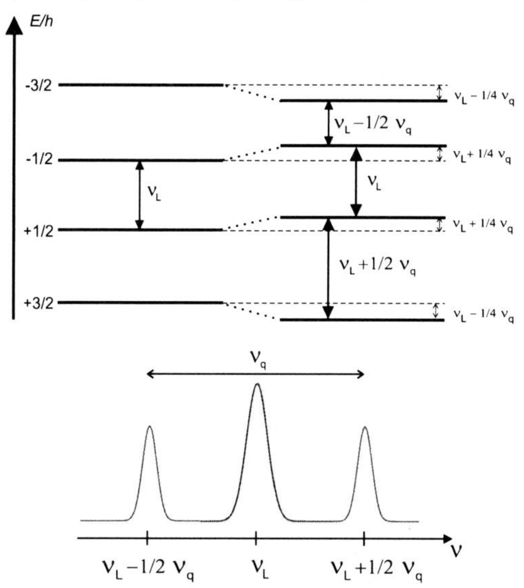


(b) Level splitting and corresponding NMR spectrum for $\theta = 90°$.

**Figure 2.2:** Level splitting due to first-order quadrupole effects in a strong magnetic field for $I = 3/2$, $\eta = 0$ and two different orientations of the EFG in the field. The left side shows the equally-split nuclear magnetic energy levels due to the Zeeman interaction, separated by the Larmor frequency. The right side shows the additional shifting due to first-order quadrupole effects. Magnetic hyperfine interactions and second-order quadrupole effects are neglected.

For $I = 3/2$ and $\theta = 0°$, the levels $m = \pm 3/2$ are shifted by $E^{(1)}_{\pm 3/2}(\theta = 0°) = +1/2\nu_q$, while the levels $m = \pm 1/2$ are shifted by $E^{(1)}_{\pm 1/2}(\theta = 0°) = -1/2\nu_q$, resulting in a splitting of the resonance into three lines, where the satellites are separated from each other by the amount $2\Delta^{(1)}(\theta = 0°) = 2\nu_q$ [see Fig. 2.2(a)]. For $I = 3/2$ and $\theta = 90°$, the $m = \pm 3/2$ levels are shifted by $E^{(1)}_{\pm 3/2}(\theta = 90°) = -1/4\nu_q$, while the $m = \pm 1/2$ levels are shifted by $E^{(1)}_{\pm 1/2}(\theta = 90°) = +1/4\nu_q$, resulting in the spacing $2\Delta^{(1)}(\theta = 90°) = \nu_q$ between both satellites [see Fig. 2.2(b)]. Note that it is better to determine the first-order quadrupole shift $\Delta^{(1)}(\theta)$ by measuring the distance $2\Delta^{(1)}(\theta)$ between two satellites instead of measuring the spacing between a satellite and the central line, since the central line might be shifted by the second-order quadrupole shift.

Concerning the second-order quadrupole shift, the prefactors $A$, $B$ and $C$ in Eq. (2.34) become simple constants in the case of uniaxial symmetry: $A(\phi, \eta) = -27/8$, $B(\phi, \eta) = 30/8$ and $C(\phi, \eta) = -3/8$ [56]. The second-order quadrupole shift of a nucleus with a nuclear spin of $I = 3/2$ is then given by [32, 56]:

$$\Delta^{(2)}\nu_m(\theta) = \frac{3\nu_q^2}{16\nu_L}(1 - \cos^2\theta)(1 - 9\cos^2\theta)\,. \tag{2.39}$$

In case of a large quadrupole frequency $\nu_q$, the second-order quadrupole shift of the central resonance line can be significant and must be subtracted from the measured frequency before evaluating the Knight shift.

### 2.2.4 Summary

Fig. 2.3 summarizes all possible interactions, the resulting shifts of the energy levels and the corresponding NMR spectra for a nuclear spin of $I = 3/2$ in a uniaxial symmetry ($\eta = 0$), a parallel orientation of the principal axis of the EFG with respect to the applied magnetic field ($\theta = 0°$, $Z = z$) and within an electronic environment with $V_{zz} \neq 0$. The degeneracy of the nuclear magnetic energy levels is lifted by applying a magnetic field $H_0$ due to the Zeeman interaction $\mathcal{H}_Z$. A delta peak at the Larmor frequency $\nu_L$ appears in the spectrum corresponding to a transition between two adjacent energy levels. The interaction between different nuclear spins $\mathcal{H}_{n-n}$ broadens the energy levels, causing a Gaussian line shape of the resonance line centered at $\nu_L$. Magnetic hyperfine interactions between the nuclear spin and the spins and the angular momenta of the electrons lead to a shift of the resonance line away from the Larmor frequency, the Knight shift $K$. Electric quadrupole interactions between the quadrupole moment of the nucleus and the surrounding EFG shift the energy levels additionally in such a way that the degeneracy of the transitions is lifted. This results in a splitting of the NMR spectrum into three lines, where in first order the central transition remains unaffected. For $\theta = 0°$, as considered in Fig. 2.3, the distance between the two satellites amounts to $2\nu_q$. In second order and for $\theta \neq 0°$, the central line shifts additionally by a factor $\Delta^{(2)}\nu(\theta)$ defined in Eq. (2.39). For $\theta = 0°$ however, as sketched in Fig. 2.3, there is no second-order quadrupole effect.

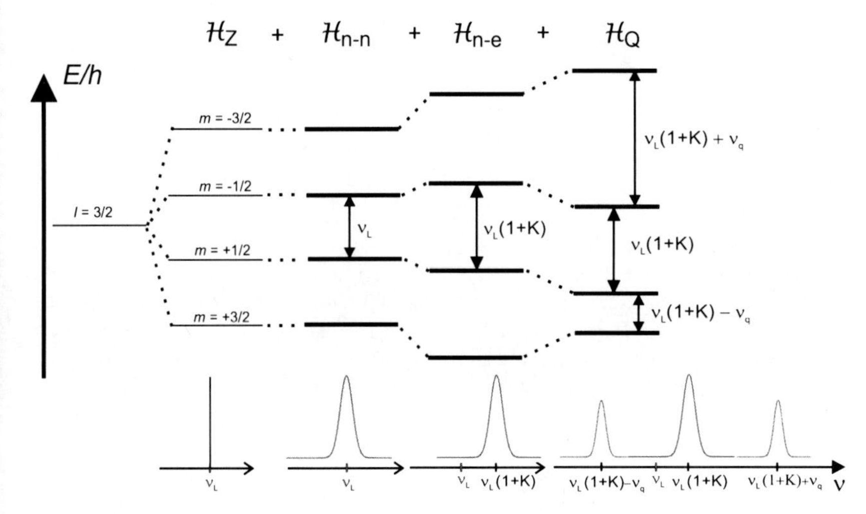


**Figure 2.3:** Summary of all possible interactions, their influence on the nuclear magnetic energy levels and the resulting NMR spectra for $I = 3/2$, $\eta = 0$, $\theta = 0°$ and $V_{zz} \neq 0$ under the condition that $\mathcal{H}_Q \ll \mathcal{H}_Z$. Note that the choice of a positive Knight shift $K$ is arbitrary. $K$ can also have negative values.

## 2.3 Dynamic Processes: Relaxation

Every magnetic moment $\vec{\mu} = \gamma\vec{I}$ exposed to a magnetic field $\vec{H}_0$ experiences a torque $\vec{T}$ acting on it:

$$\vec{T} = \frac{d\vec{I}}{dt} = \vec{\mu} \times \vec{H}_0 \,. \tag{2.40}$$

For the nuclear magnetization $\vec{M}$, defined by Eq. (3.21), the equation of motion results in:

$$\frac{d\vec{M}(t)}{dt} = \gamma(\vec{M}(t) \times \vec{H}_0) \,. \tag{2.41}$$

Its solution is a precession of the magnetization $\vec{M}$ around the magnetic field $\vec{H}_0$ with the precession frequency $\omega_L = \gamma H_0$. At thermodynamic equilibrium, described by the Boltzmann statistics, and presuming that $\vec{H}_0$ points along the $z$-axis of the laboratory system, the components of the nuclear magnetization are: $M_z = M_0$ and $M_y = M_x = 0$. The thermodynamic equilibrium can be destroyed by applying a second, circularly polarized field $\vec{H}_1$ with the rotation frequency $\omega$ perpendicular to $\vec{H}_0$. The total magnetic field $\vec{H} = \vec{H}_0 + \vec{H}_1$ acting on the magnetization during the application of the perpendicular field $\vec{H}_1$ then becomes:

$$\begin{aligned} H_x &= H_1 \cos \omega t \\ H_y &= H_1 \sin \omega t \\ H_z &= H_0 . \end{aligned} \tag{2.42}$$

For further considerations it is convenient to go from the fixed laboratory reference frame $\mathcal{S}$ to a reference frame $\mathcal{S}'$ which rotates with the frequency $\omega$ of the high frequency field $\vec{H}_1$ around the direction $z = z'$ of the applied magnetic field $\vec{H}_0$. The time dependence of $\vec{H}_1(t)$ disappears in $\mathcal{S}'$, since both are rotating with the same frequency $\omega$. The direction of $\vec{H}_1$ can be chosen with arbitrary phase in $\mathcal{S}'$, perpendicular to $H_0$. In the following it will be considered that $\vec{H}_1$ is fixed along $\vec{e}_{x'}$. During the pulse duration $t_p$ of the radio frequency field the effective magnetic field in $\mathcal{S}'$ is then given by:

$$\vec{H}_{eff} = (H_0 + \omega/\gamma)\vec{e}_{z'} + H_1\vec{e}_{x'} . \tag{2.43}$$

The magnetization will process around this effective field (see Fig. 2.4). In the case of resonance, where $\omega = -\omega_L = -\gamma H_0$, the effective field is $\vec{H}_{eff} = \vec{H}_1$ and the magnetization precesses around $\vec{H}_1$ (see Fig. 2.5). This precession is much slower than the one around $\vec{H}_0$ in the laboratory frame, since its precession frequency is $\omega' = -\gamma H_1$ and $H_1 \ll H_0$.[11]
The angle of tilt of the magnetization is given by:

$$\alpha = \omega' t_p = -\gamma H_1 t_p . \tag{2.44}$$

By carefully choosing the pulse duration time $t_p$ and the field strength $H_1$, it is in particular possible to flip the magnetization into the $y'$- or $-z$-direction. These are called 90° and 180°-pulse, respectively, and play an important role in the determination of the relaxation times $T_1$ and $T_2$, defined below.

Each angle of tilt (besides a 180° pulse which only inverts $M_z$) results in a reduction of $M_z$ and some non-vanishing components $M_x$ and $M_y$. After the disturbance by the radio frequency pulse, the system will relax back to its thermodynamic equilibrium $M_z = M_0$ and $M_x = M_y = 0$. The decay of the transverse components $M_x$ and $M_y$ and the recovery of the longitudinal component $M_z$ back to its equilibrium value $M_0$ can be described by two relaxation times: the transverse relaxation time $T_2$, and the longitudinal relaxation time $T_1$. The relaxation equations in the laboratory system $\mathcal{S}$ are:

$$\begin{aligned} \frac{dM_x(t)}{dt} &= -\frac{M_x(t)}{T_2} \\ \frac{dM_y(t)}{dt} &= -\frac{M_y(t)}{T_2} \\ \frac{dM_z(t)}{dt} &= \frac{M_0 - M_z(t)}{T_1} . \end{aligned} \tag{2.45}$$

[11] Experimentally, the circularly polarized field $\vec{H}_1$ is realized by applying high frequency pulses generating an alternating magnetic field $\vec{H}_{rf}(t)$ perpendicular to $H_0$. This field can be decomposed into two contra-rotating components with the frequencies $\omega$ and $-\omega$. The component rotating with $-\omega$ can be neglected, since it is far away from the effective resonance. It is thus sufficient to deal with one magnetic field $\vec{H}_1(t)$ rotating with $\omega$.

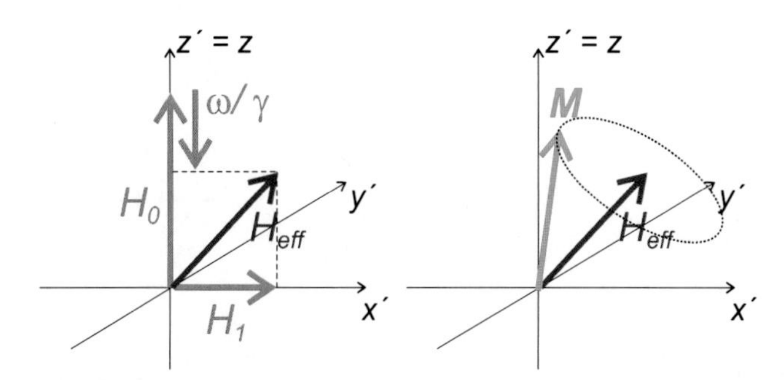


**Figure 2.4:** Sketch of the effective field in the rotating reference frame $\mathcal{S}'$ (left panel) and the precession of the magnetization around the effective field in $\mathcal{S}'$ (right panel) in a non-resonant case.

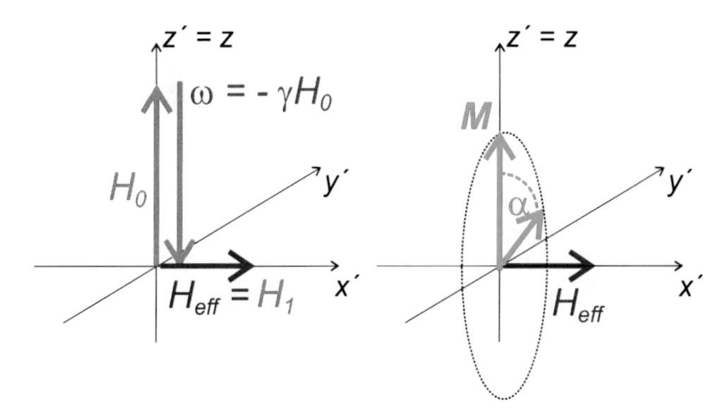


**Figure 2.5:** Sketch of the effective field in the rotating reference frame $\mathcal{S}'$ in case of resonance $\omega = -\omega_L = -\gamma H_0$ (left panel). The effective field equals $H_1$ and lies along $x'$. The magnetization precesses with the angle of rotation $\alpha$ around the effective field (right panel). For $\alpha = 90°$ (180°) it will be aligned along $y'$ $(-z')$ z' $z'$.

$T_1$ is also called spin-lattice relaxation time, since for the recovery of $M_z$ to $M_0$ a redistribution of the nuclear magnetic energy levels is needed, which can only occur if the nuclear spin system exchanges energy with the lattice (more precisely with excitations of the lattice such as phonons, spinons and so on). In contrast, for the relaxation of $M_x$ and $M_y$ back to zero due to the dephasing of the nuclear spins no energy transfer is needed. This dephasing is caused by inhomogeneities of the external magnetic field and by local fields from electronic spins as well as by local fields stemming from interactions between the nuclear spins. On account of this $T_2$ is also called spin-spin relaxation time. Putting

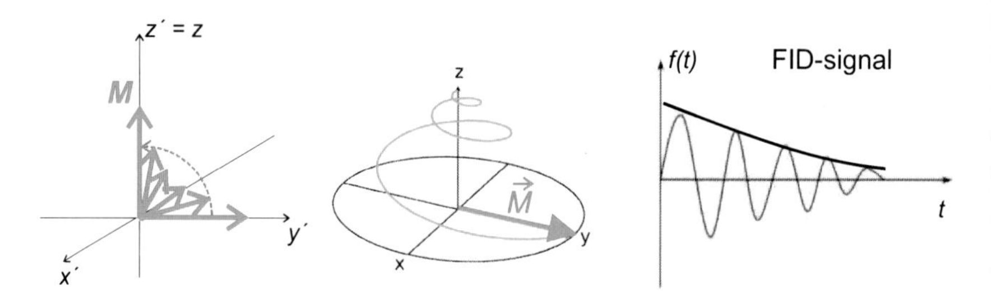


**Figure 2.6:** Relaxation of the nuclear magnetization $\vec{M}$ back to its equilibrium value $M_z = M_0$ after being disturbed by a 90° pulse, flipping it into the $x'y'$ ($xy$)-plane in the rotating frame $\mathcal{S}'$ (left panel) and the laboratory system $\mathcal{S}$ (central panel, adopted from [36]). The relaxation (here the effect of $T_1$ is exaggerated versus the effect of $T_2$) is accompanied by a precession of $\vec{M}(t)$ around the $z$-axis in the laboratory system. The right panel shows the free induction decay (FID) signal, induced by the precession of non-zero $M_x$ and $M_y$ components of the nuclear magnetization.

the relaxation equations (2.45) together with Eq. (2.41) describing the precession of $\vec{M}$ around $\vec{H}$, one arrives at the Bloch equations (in the laboratory frame $\mathcal{S}$) [61]:

$$\begin{aligned}
\frac{dM_x(t)}{dt} &= \gamma(\vec{M}(t) \times \vec{H})_x - \frac{M_x(t)}{T_2} \\
\frac{dM_y(t)}{dt} &= \gamma(\vec{M}(t) \times \vec{H})_y - \frac{M_y(t)}{T_2} \\
\frac{dM_z(t)}{dt} &= \gamma(\vec{M}(t) \times \vec{H})_z + \frac{M_0 - M_z(t)}{T_1} .
\end{aligned} \tag{2.46}$$

These equations describe how the magnetization relaxes back to its equilibrium and simultaneously precesses freely around $z$. An example is given in Fig. (2.6) for a magnetization which has been flipped into the $xy$-plane. For the solution of Eq. (2.46), for which it is convenient to return to the rotating frame system, the reader is referred to [28].

## 2.3.1 How to Measure Nuclear Magnetism?

The rotation of the in-plane magnetization $M_x$ and $M_y$ results in a time-dependent magnetic flux through the locally-fixed receiver solenoid[12]. This flux induces a small alternating voltage in the coil which can be detected[13]. The detected voltage will oscillate to zero, reflecting the decay of the transverse magnetization due to spin-spin relaxation (see Section 2.3.3 for a detailed discussion of possible spin-spin relaxation mechanisms). This is called **Free Induction Decay (FID)**. It is sketched in the right panel of Fig. 2.6. The FID is modulated in accordance with the slightly different Larmor frequencies of the

[12] The receiving and emitting coils are usually one and the same solenoid. The separation between in- and out-coming signals is realized by the use of a $\lambda/4$-cable, explained in more detail in Chapter 5.

[13] Typical values of the induced voltage are of the order of a few microvolts. An amplifier is needed to detect these small voltages (see Chapter 5).

nuclear spins, stemming from local inhomogeneities of the magnetic field. Performing a Fourier transformation of the FID signal reveals a fraction of the NMR spectrum centered around the irradiation frequency $\nu_{irr}$. If the resonance lines are narrow and only slightly shifted with respect to each other, one can even observe a whole NMR spectrum by measuring with one irradiation frequency (see Section 8.4 for an example). This simultaneous observation of different NMR lines is an advantage of pulsed NMR compared to continuous wave NMR (CW NMR). However, the interval $[\nu_{irr} - \Delta\nu; \nu_{irr} + \Delta\nu]$ of measurable frequencies, which one can observe by applying a pulse with the irradiation frequency $\nu_{irr}$ is limited by the chosen pulse length $t_p$. It is roughly given by the FWHM of its Fourier transformation. The Fourier transformation of a rectangular pulse with pulse width $t_p$ reads [35]:

$$F(\nu) = F_0 \frac{\sin[\pi(\nu - \nu_{irr})t_p]}{\pi(\nu - \nu_{irr})t_p} \,. \tag{2.47}$$

It is sketched in Fig. 2.7 and gives

$$\Delta\nu \approx 1/t_p \,. \tag{2.48}$$

If the frequency range to be measured is broader, a series of measurements with different pulse frequencies $\nu$ (frequency sweep) or alternatively a sweep of the applied magnetic field value is necessary to obtain the whole NMR spectrum. Accordingly, measurements of $T_1$ and $T_2$ may have to be done at different positions in the spectrum. Alternatively the pulse width can be changed by adjusting the magnitude of the oscillating field $H_1$, using an adjustable attenuator. One has to be careful to keep the angle of tilt $\alpha$ at the desired value (mainly 90° or 180°, according to Eq. (2.44)).

Another advantage of pulsed NMR in comparison to CW NMR is the fact that the use of pulses allows the design of different pulse sequences for specific measurements, such as $T_1$ and $T_2$ measurements. For instance the NMR spectrum itself is not just obtained by Fourier transforming the FID, but by applying a special pulse sequence, the so-called

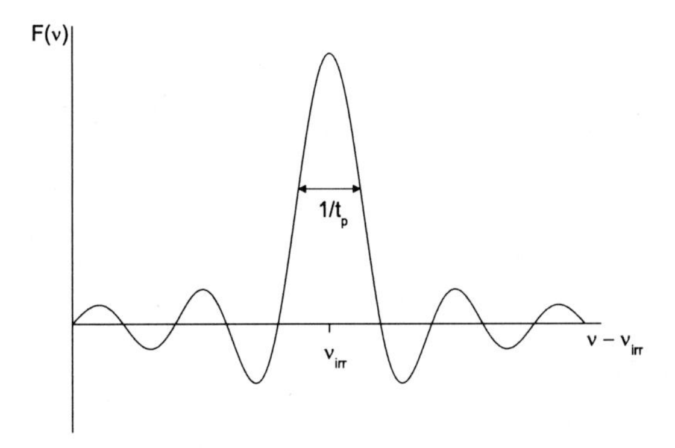


**Figure 2.7:** Fourier transformation of a rectangular pulse with a pulse width (pulse duration time) $t_p$ according to Eq. (2.47). Its FWHM of $1/t_p$ defines the observable frequency range [see Eq. (2.48)].

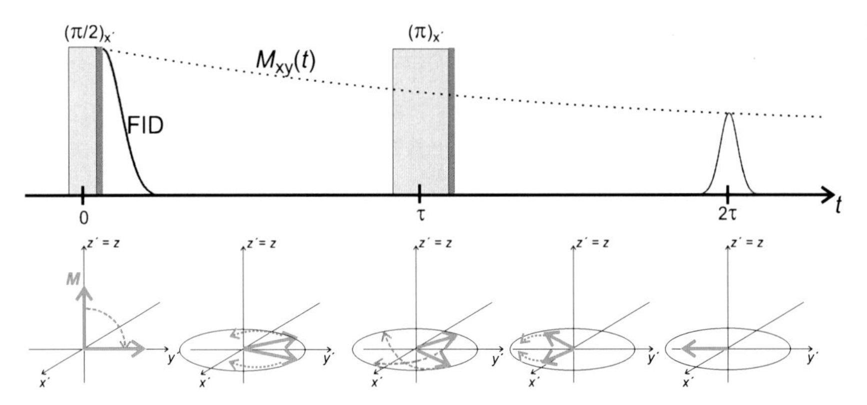


**Figure 2.8:** Sketch of the Hahn spin echo pulse sequence on a time scale (upper panel) and the corresponding movement of the nuclear magnetization in the rotating frame system $\mathcal{S}'$ (lower panel). The dead time of the detection electronics and the fall time of the pulses are sketched together in light red.

Hahn spin echo. It is depicted in Fig. 2.8. A first 90°-pulse applied along the $x'$-axis flips the magnetization from its equilibrium position into the $x'y'$-plane. Due to spin-spin interactions and local field inhomogeneities the individual moments rotate with different speeds within the $x'y'$ plane and thus begin to dephase. This induces the FID signal in the coil. Subsequently the transverse magnetization $M_{xy}$ decreases. After a time $\tau$ a 180° pulse is applied to the system, again along $x'$. It flips the different components of the transverse magnetization within the $x'y'$-plane, such that they will refocus at the time $2\tau$ and give rise to the spin echo, whose intensity is related to the magnitude of $M_{xy}(2\tau)$. There are two big advantages of the Hahn spin echo versus the direct observation of the FID. Firstly, the influences of local field inhomogeneities, which are static on the time scale of the measurement, cancel out due to the refocussing. Only intrinsic effects of spin-spin interactions remain as a source for the reduction of $M_{xy}(2\tau)$. The FID decays with the time $T_2^*$, containing both intrinsic spin-spin interactions and the effect of inhomogeneities of the static magnetic field (see Section 2.3.3). On the contrary, the intensity of the Hahn spin echo can be used to measure the spin-spin relaxation time $T_2$. Secondly, since the spin echo appears at a time distance $\tau$ away from the last pulse, it is not superimposed by possible tails of the pulses during their fall time and its observation is not hindered by the dead time of the detection electronics.

## 2.3.2 Spin-Lattice Relaxation

### 2.3.2.1 Sources of Spin-Lattice Relaxation

As described in detail in Appendix A.2, both fluctuating magnetic fields $\vec{h}(t)$ and fluctuating components of the EFG $V_k(t)$ may induce spin-lattice relaxation processes, comprised in the time-dependent Hamiltonian $\mathcal{H}_1(t)$ defined in Eq. (A.20). The fluctuating components of the EFG may be produced by phononic excitations or charge fluctuations, e.g. the motion of holes in the spin ladder system $Sr_{14}Cu_{24}O_{41}$ [49]. Fluctuating magnetic fields may arise from nuclear spin-spin interactions, interactions with conduction electrons or spin excitations of localized electronic spins.

The time dependence of the fluctuating magnetic field $\vec{h}(t)$ is described by its autocorrelation function at the times $t$ and $t+\tau$ [62, 63]:

$$G_{\alpha\alpha}(\tau) = \langle h_\alpha(t) h_\alpha(t+\tau)\rangle_t = \langle h_\alpha^2\rangle \exp\left(-|\tau|/\tau_c\right) , \tag{2.49}$$

with $\alpha = x, y, z$. The last relation assumes that the three field components fluctuate independently from each other and that their autocorrelation functions decrease exponentially with the same characteristic correlation time $\tau_c$. According to Eqs. (A.21) and (A.24) the spin-lattice relaxation rate is proportional to the spectral density of the fluctuating field components:

$$J_{\alpha\alpha}(\omega) = \frac{\gamma^2}{2}\int_{-\infty}^{\infty} \langle h_\alpha(t) h_\alpha(t+\tau)\rangle_t \exp(-i\omega\tau)d\tau\,. \tag{2.50}$$

$J_{\alpha\alpha}(\omega)$ has usually a Lorentzian shape. Applying the density matrix formalism [28], which connects the time-dependent populations of the eigenstates of the static Hamiltonian $\mathcal{H}_0$ to the time-dependent perturbing Hamiltonian $\mathcal{H}_1(t)$ and assuming an orientation of the applied magnetic field $H_0$ parallel to the $z$-axis, one obtains [28]:

$$\frac{1}{T_1} = \gamma^2(\langle h_x^2\rangle + \langle h_y^2\rangle)\frac{\tau_c}{1+\omega_L^2\tau_c^2}\,. \tag{2.51}$$

It is obvious from Eq. (2.51) that only fluctuations perpendicular to the externally applied magnetic field lead to spin-lattice relaxation processes. $T_1^{-1}$ has a maximum for $\omega_L\tau_c = 1$, where the characteristic frequency of the fluctuations, $\tau_c^{-1}$, matches the Larmor frequency $\omega_L$. For $\omega_L\tau_c \ll 1$ and $\omega_L\tau_c \gg 1$ the spin-lattice relaxation rate decreases again (see Fig. 2.10). The autocorrelation function (2.49) can basically describe any fluctuating local magnetic field. In the original paper introducing this so-called BPP[14]-model, the local fluctuating field was assumed to stem from the Brownian motion of molecules in liquids and gases and $\tau_c$ was related to the viscosity of the liquids and the molecular dimensions [62, 63]. Equation (2.49) can also describe spin fluctuations in magnetic materials. In both cases, the effective correlation time (of diffusing atoms or fluctuating electronic spins) shows an activated temperature dependence of the form [27, 64–66]:

$$\tau_c = \tau_0 \exp(E_a/k_B T)\,, \tag{2.52}$$

where $E_a$ is the activation energy of the ionic motion (or of the electronic spin fluctuations, respectively) and $\tau_0$ is the correlation time at high temperatures.

Based on the spectral density for fluctuating magnetic fields [Eq. (2.50)] and expressing the internal magnetic field at the nucleus produced by an electron corresponding to the Fermi contact part of Eq. (2.17) (which is the most relevant part for $s$-electron metals), one arrives at the following relation for the spin-lattice relaxation in correlated metals [67]:

$$\frac{1}{T_1} = \frac{2\gamma^2 k_B T}{g^2\mu_B^2}\sum_{\vec{q}} |A_\perp(\vec{q})|^2 \frac{\chi_\perp''(\vec{q},\omega_L)}{\omega_L}\,. \tag{2.53}$$

[14] BPP= Bloembergen, Purcell and Pound.

Here, $g$ is the Landé factor, $A_\perp(\vec{q})$ are the transverse components of the magnetic hyperfine coupling and $\chi''_\perp(\vec{q},\omega_L)$ is the transverse imaginary part of the dynamic susceptibility $\chi(\vec{q},\omega_L) = \chi'(\vec{q},\omega_L)+i\chi''(\vec{q},\omega_L)$ at the wave vector $\vec{q}$ and the Larmor frequency $\omega_L$. Hence, the problem of calculating $T_1^{-1}$ in metals (and correspondingly in magnetic materials) depends on the knowledge of the hyperfine coupling $A_\perp(\vec{q})$ and the dynamic susceptibility $\chi''_\perp(\vec{q})$. The $\vec{q}$-dependence of the hyperfine coupling reflects that not every electronic spin fluctuation with an arbitrary wavevector may induce a flip of a nuclear spin. Depending on the nuclear site under consideration, $A_\perp(\vec{q})$ can have different $\vec{q}$ dependencies which act as a filter for certain regions of the Brillouin zone and may conceal important fluctuations such as antiferromagnetic fluctuations at $(\pi,\pi)$ or ferromagnetic fluctuations at $(0,0)$.

#### 2.3.2.2 Inversion Recovery Method to Measure $T_1^{-1}$

There are several possible pulse sequences to measure the spin-lattice relaxation rate $T_1^{-1}$. In the course of this thesis, only the *inversion recovery method* was used, which will be described in the following. It consists of a $\pi$ pulse followed by a normal Hahn spin echo pulse sequence after a variable delay time $\tau_1$. The pulse sequence and the corresponding movement of the magnetization in the rotating frame are depicted in Fig. 2.9 for short and long delay times $\tau_1$. The first 180° pulse flips the magnetization into the $-z'$-direction (it inverses the populations of the nuclear magnetic energy levels). Due to spin-lattice interactions, the magnetization begins to relax back to its equilibrium value during the time $\tau_1$. The following Hahn spin echo sequence measures the longitudinal magnetization $M_z(\tau_1)$ as a function of the delay time $\tau_1$ (see Fig. 2.8 and Section 2.3.1 for the explanation of the Hahn spin echo pulse sequence).

If $\tau_1$ is short [Fig. 2.9(a)], the magnetization cannot relax much and the following Hahn spin echo will measure a negative magnetization. If $\tau_1$ is long [Fig. 2.9(b)], the longitudinal magnetization can relax a lot and will already point along $+z'$ before the Hahn spin echo pulse sequence is applied. A complete inversion recovery measurement consists of a series of such pulse sequences, where $\tau_1$ is gradually increased and the magnetization is measured as a function of $\tau_1$. The resulting function $M_z(\tau_1)$ [Fig. 2.9(c)] allows to extract $T_1$ by fitting $M_z(\tau_1)$ with a (multi-) exponential relaxation function.

#### 2.3.2.3 Spin-Lattice Relaxation Functions

The relaxation function describing the recovery of the longitudinal nuclear magnetization $M_z(t)$ depends on several details which are:

- the value of the nuclear spin $I$
- the transition $(m) \leftrightarrow (n)$ on which the recovery is measured
- the measurement method (inversion recovery, saturation recovery, ...)
- the ratio between the Zeeman term $\mathcal{H}_Z$ and the quadrupole contribution $\mathcal{H}_Q$ of the static Hamiltonian

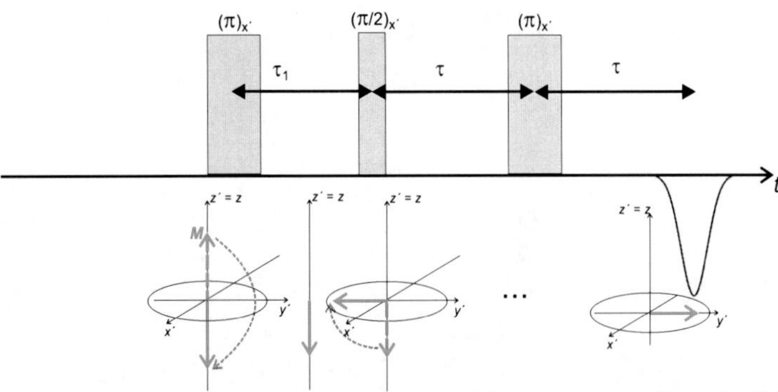


(a) Short delay times $\tau_1$ prevent a strong relaxation of the magnetization. Hence the Hahn spin echo sequence $\pi/2 - \tau - \pi - \tau$ measures a negative magnetization $M_z(\tau_1)$.

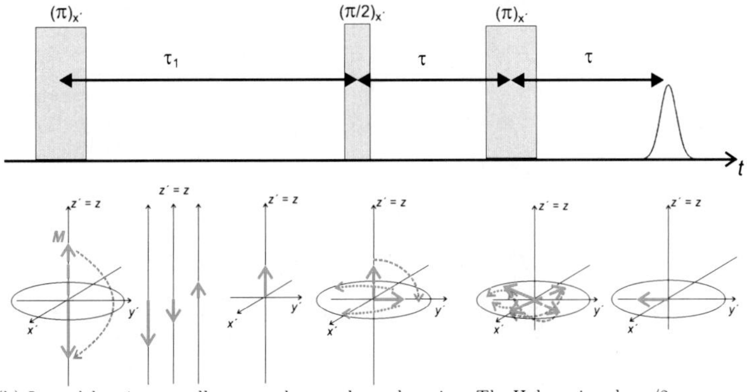


(b) Long delay times $\tau_1$ allow a nearly complete relaxation. The Hahn spin echo $\pi/2-\tau-\pi-\tau$ measures a positive $M_z(\tau_1)$.

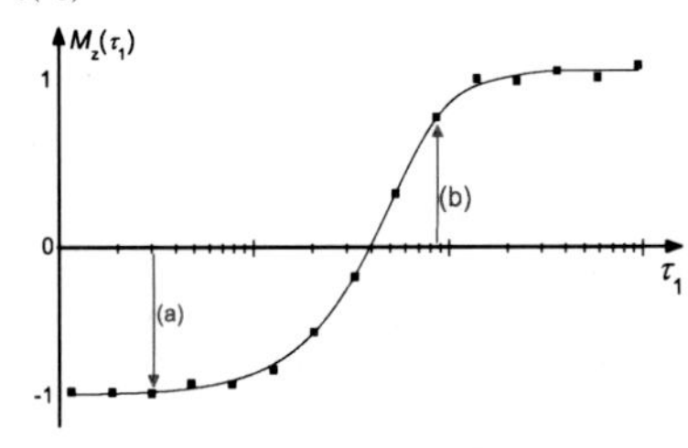


(c) Example of an inversion recovery measurement on $LaO_{0.9}F_{0.1}FeAs_{1-\delta}$ at $T = 250\,K$. $M_z(\tau_1)$ changes from negative to positive values as $\tau_1$ increases and saturates for $\tau_1 \gtrsim T_1$. Black squares are data points, the solid line is a fit of Eq. (2.54) to extract $T_1$. Red arrows denote the situation plotted in (a) and (b), respectively.

**Figure 2.9:** Sketch of the pulse sequence and the corresponding movement of $M_z$ in the rotating frame system for the inversion recovery method to measure $T_1$ for (a) a short delay time $\tau_1$ and (b) a long delay time $\tau_1$. (c) shows an example of the development of $M_z(\tau_1)$.

- the relaxation mechanism (magnetic, quadrupolar, or mixed)
- the definition of the spin-lattice relaxation rate $T_1^{-1}$ relative to the transition probabilities $W_{mn}$

A detailed derivation of the spin-lattice relaxation function $M_z(t)$ referring to all these points is recapitulated in Appendix A.2. In this Chapter, only the the most important case for this work, **the purely magnetic relaxation of a nuclear spin $I = 3/2$**, will be considered. It is assumed that the Zeeman term of the static Hamiltonian is the leading interaction, which is only slightly perturbed by the quadrupole coupling: $\mathcal{H}_Q \ll \mathcal{H}_Z$. Measured at the central resonance line $(m = +\frac{1}{2}) \leftrightarrow (m = -\frac{1}{2})$, the relaxation is described by [68–72]:

$$M_z(\tau_1) = M_0\{1 - f[0.9\mathrm{e}^{-6\tau_1/\mathrm{T}_1} + 0.1\mathrm{e}^{-\tau_1/\mathrm{T}_1}]\}\,. \tag{2.54}$$

Measured at the satellites $(m = +\frac{3}{2}) \leftrightarrow (m = +\frac{1}{2})$ and $(m = -\frac{1}{2}) \leftrightarrow (m = -\frac{3}{2})$, the magnetic relaxation has to be fitted with [68–72]:

$$M_z(\tau_1) = M_0\{1 - f[0.4\mathrm{e}^{-6\tau_1/\mathrm{T}_1} + 0.5\mathrm{e}^{-3\tau_1/\mathrm{T}_1} + 0.1\mathrm{e}^{-\tau_1/\mathrm{T}_1}]\}\,. \tag{2.55}$$

If the quadrupole splitting is very small and all three NMR lines (central line and satellites) are excited simultaneously by the applied radio frequency pulse, the relaxation function becomes [70, 72]:

$$M_z(\tau_1) = M_0\{1 - f\mathrm{e}^{-\tau_1/\mathrm{T}_1}\}\,, \tag{2.56}$$

while in NQR experiments the relaxation obeys[15,16] [74]:

$$M_z(\tau_1) = M_0\{1 - f\mathrm{e}^{-3\tau_1/\mathrm{T}_1}\}\,. \tag{2.57}$$

The prefactor $f$ in Eqs. (2.54) to (2.57) is the inversion fraction. It describes the completeness of the inversion. For a complete inversion of $M_z$ in an inversion recovery measurement $f$ equals 2. A complete inversion is usually hard to achieve and $f$ will usually lie between 1.7 and 1.9. Sometimes a distribution of spin-lattice relaxation times around a characteristic $T_1$ has to be considered. In this case the exponents of Eqs. (2.54) to (2.57) have to be raised to the power of a stretching exponent $\lambda$, with $0 < \lambda \leqslant 1$. For a detailed discussion of the use of the stretching exponent $\lambda$ the reader is referred to Appendix A.3.

#### 2.3.2.4 Korringa Relation

In metals with non-interacting electrons (Fermi gas) and the Fermi contact interaction as the main source for hyperfine interactions between the electron and the nuclear spin (as it is the case in mono-valent metals) one can show that [75]:

$$K_s^2 T_1 T = \frac{\hbar}{4\pi k_B}\left(\frac{\gamma_e}{\gamma_n}\right)^2 = S_0 = \text{const.} \tag{2.58}$$

[15] The relaxation function (2.57) holds for cases of axially-symmetric field gradients ($\eta = 0$). The more complicated situation of $\eta \neq 0$ has been considered by Chepin and Ross [73].

[16] Note that the spin-lattice relaxation time $T_1$ can be arbitrarily defined with relation to the transition probability $W_1$ (see App. A.2 for details). In the previous formulas, $T_1$ was chosen according to Eq. (A.29) as $T_1^{-1} = 2W_1$. In the case of Eq. (2.57), $T_1$ is defined as $T_1^{-1} = (2/3)W_1$ [72].

This is the so-called Korringa relation, where the Korringa constant $S_0$ depends only on the probed isotope (through $\gamma_n$). Experimentally however, only very few materials fulfill Eq. (2.58). For most materials correlations between electrons have to be taken into account [67, 75–77]. These include electron-electron exchange interactions (Stoner enhancement) as well as other possible magnetic correlations between conduction electrons. It is therefore common to use the modified Korringa ratio [36, 77–81]:

$$K_s^2 T_1 T = \alpha S_0 \, . \tag{2.59}$$

$\alpha = 1$ reflects the behavior of a normal metal without any correlations (Fermi gas). The deviations from $\alpha = 1$ are a measure of the strength of possible correlations (Fermi liquid). $\alpha > 1$ presents the tendency of the observed system towards ferromagnetic correlations at $\vec{q} = 0$ (e.g. Stoner enhancement of the uniform susceptibility) [36, 67, 76, 77], and $\alpha < 1$ depicts its tendency towards antiferromagnetic correlations at $\vec{q} \neq 0$ [36, 77].

In the case of high temperature superconductors it is common to ascribe the behavior $K_s^2 T_1 T = \alpha S_0 =$ const. in the normal state to a Fermi liquid character of the superconductor [23, 34, 78].

However, one has to be careful when doing a quantitative interpretation based on the value of $\alpha$. First, it was shown that also disorder may enhance the Korringa ratio $\alpha$ [77]. Second, hyperfine form factor effects can also radically change the value of $\alpha$, by filtering out important regions in $\vec{q}$-space (see e.g. [82]).

Even the interpretation of a constant or not-constant Korringa ratio has to be examined carefully. Since both the Knight shift $K_s$ and the spin-lattice relaxation rate $T_1^{-1}$ depend on the density of states of electrons at the Fermi level, a temperature- (or field-) dependent density of states, for instance due to electron electron interactions, may also lead to a temperature- (or field-) dependent Korringa ratio [75]. This does not exclude that the system can be described within a Fermi liquid picture. The observation of a constant Korringa relation is thus a proof of a Fermi liquid, but its absence is not a strict counter evidence against it. Furthermore, the absolute value of $\alpha$ should be interpreted with great care.

### 2.3.3 Spin-Spin Relaxation

Spin-spin relaxation is caused by both direct dipolar and indirect couplings between nuclear spins as described in Section 2.2.1 as well as by inhomogeneities of the static magnetic field. Locally fluctuating magnetic fields, which induce transitions between different energy levels and therefore spin-lattice relaxation, also cause spin-spin relaxation. This is the so-called Redfield contribution [83]. All mentioned effects cause the loss of coherence between the spins and thus a decay of the transverse components $M_x$ and $M_y$ of the nuclear magnetization. They are described by the relaxation rate $1/T_2^*$ [36]:

$$\begin{aligned} \frac{1}{T_2^*} &= \frac{1}{T_2} + \frac{1}{T_2^{inhomog}} \\ &= \frac{1}{T_2^{dip}} + \frac{1}{T_2^{indirect}} + \frac{1}{T_2^{Redfield}} + \frac{1}{T_2^{inhomog}} \, . \end{aligned} \tag{2.60}$$

The spin-spin relaxation rate $1/T_2$ is defined as the sum of the first three contributions. The last contribution is not included in $1/T_2$ since it does not stem from "real" relaxation processes but from local inhomogeneities of the static magnetic field, which lead to a dephasing due to slightly different Larmor frequencies $\omega_L$. The FID depicted in Figs. 2.6 and 2.8 decays with the relaxation time $T_2^* < T_2$. For a measurement of the spin-spin relaxation time $T_2$ one takes advantage of the Hahn spin echo pulse sequence (Fig. 2.8), where the contributions of static local field inhomogeneities cancel out due to the refocussing of the different components of the transverse magnetization after the 180° pulse.[17] In order to measure the spin-spin relaxation time $T_2$ one observes the decay of the Hahn spin echo as a function of the time $\tau$ between the two pulses (see Fig. 2.8). An exponential fit to the decay of $M_{xy}(2\tau)$ yields the spin-spin relaxation rate $1/T_2$:

$$M_{xy}(2\tau) = M_0 \exp(-2\tau/T_2)\,. \tag{2.61}$$

A more elaborate fit discriminates between the already discussed decay rate due to spin-lattice processes, $1/T_2^{Redfield}$, and the decay rate due to nuclear spin-spin interactions, $1/T_{2G}$ [49]:

$$M_{xy}(2\tau) = M_0 \exp\left[-2\tau/T_2^{Redfield} - (2\tau)^2/(2T_{2G}^2)\right]\,. \tag{2.62}$$

$1/T_2^{Redfield}$ can be determined from $1/T_1$ [84, 85]. The remaining fit parameter $1/T_{2G}$ comprises both the direct dipolar contribution $1/T_2^{dip}$ and the indirect coupling between nuclear spins $1/T_2^{indirect}$. The latter contains information about the static, $\vec{q}$-dependent susceptibility $\chi'(\vec{q})$ complementary to $1/T_1$ [46, 48, 49] (see discussion in Section 2.2.1).

Following the approach discussed in Section 2.3.2.1, the spin-spin relaxation rate due to fluctuating magnetic fields (Redfield contribution) can be calculated to [28, 35]:

$$\frac{1}{T_2^{Redfield}} = \gamma^2\left[\langle h_z^2\rangle\tau_c + \frac{1}{2}(\langle h_x^2\rangle + \langle h_y^2\rangle)\frac{\tau_c}{1+\omega_L^2\tau_c^2}\right] = \frac{1}{T_2'} + \frac{1}{2T_1}\,. \tag{2.63}$$

This expression contains two contributions. The first one, $1/T_2'$, represents the dephasing of the spins due to the spread in fluctuating fields parallel to the applied magnetic field $H_0$ (secular broadening). The second contribution, $1/2T_1$, reflects a spread in the transverse components of the fluctuating field stemming from the lifetime broadening of the nuclear energy levels. It is related to $1/T_1$. Fig. 2.10 summarizes the dependences of $1/2T_1$ and $1/T_2'$ on the correlation time $\tau_c$. Whereas the longitudinal contribution $1/2T_1$ goes through a maximum at $\tau_c = 1/\omega_L$, the transverse part $1/T_2'$ increases with $\tau_c$ up to $\tau_c = 1/\gamma h_z$. For larger values of $\tau_c$ the Redfield theory cannot be applied any more [28].

In this work the spin-spin relaxation will not be discussed explicitly. It is however important to have a rough estimate of $T_2$ for the choice of appropriate pulse sequence parameters.

[17] Local field inhomogeneities may arise due to inhomogeneities of the external field as well as due to internal dipole fields or inhomogeneities of local fields produced by electronic spins. All these contributions will be eliminated by the Hahn spin echo pulse sequence, as long as they are static over the time scale of the measurement [27], which means as long as $\tau_c \gg T_2$.

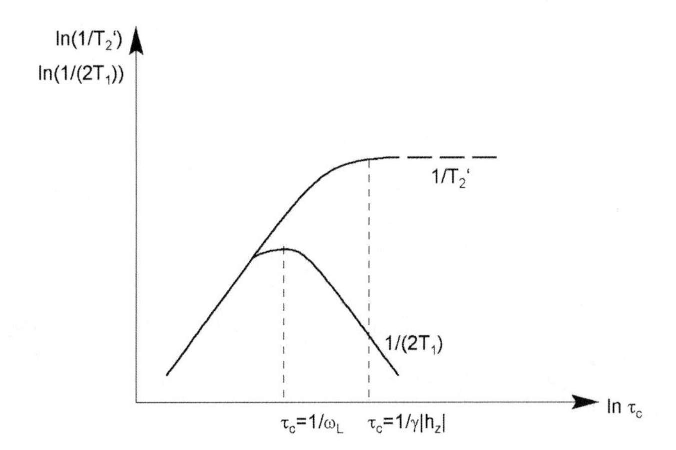


**Figure 2.10:** Relaxation rates $1/T_2'$ and $1/2T_1$ as a function of the correlation time $\tau_c$ according to Equations (2.51) and (2.63) on a double logarithmic scale [28] (graph adopted from [35]). The dashed line corresponds to the region $\tau_c > 1/\gamma h_z$ for which $1/T_2'$ cannot be described any more by the Redfield theory (see text).

## 2.4 Line Shape II

In addition to the effects discussed in Section 2.2.1, many other reasons which are unrelated to nuclear spin-spin interactions can influence the line shape, in most cases leading to an inhomogeneous broadening. One factor, which should and can be avoided, is certainly the inhomogeneity of the applied static magnetic field. This effect can be diminished by using high resolution NMR magnets and small samples, which cover less space in the magnetic field and thus experience smaller gradients.

Spin-lattice relaxation processes lead to an additional broadening of all resonance lines with a factor of $\hbar/T_1$ (*lifetime broadening* due to the Heisenberg uncertainty principle, see also discussion of Eq. (2.63)).

Inhomogeneities of the internal magnetic field due to impurities, vacancies, spin density oscillations, static local fields and so on lead to a distribution of Knight shifts $K$ and thus also to a broadening of the resonance line. The distribution of Knight shifts might thereby stem from a distribution of hyperfine couplings or from inhomogeneities of the local spin susceptibility itself [see Eq. (2.21)]. The corresponding line broadening affects both the central line and the satellites in the same way in first order [86].

In crystals with a non-vanishing EFG, imperfections such as dislocations, vacancies, interstitial sites, strains, foreign atoms as well as intrinsic charge inhomogeneities (charge density wave, e.g.), will cause a variation of the intensity and orientation of the EFG from site to site. In first order of perturbation of the Hamiltonian, this local variation of the EFG leads to a broadening (and in some cases even to a complete blurring) of the NMR satellite lines, while the NMR central resonance line is only affected in second order. Correspondingly, the influence of quadrupole effects can be directly measured by observing the NQR resonance line(s).

In general, in the paramagnetic state, effects of disorder are much stronger on the quadrupole broadening than on the magnetic broadening. Any small deviation from a homogeneous charge distribution or any lattice anomaly or defect will lead to a distribution of electric field gradients and thus contribute to the quadrupole linewidth. For the discussion of the quality of a crystal it is therefore convenient to check the quadrupole linewidth (either in NQR directly or by comparing the linewidth of the NMR satellites with that of the central NMR resonance line). This will be important for the discussion of the properties of the samples presented in this thesis, above all for the characterization of LiFeAs in Section 5.3.2.

NMR spectra of powder samples will be inhomogeneously broadened due to anisotropies in the quadrupole and magnetic interactions (anisotropic Knight shift) resulting from crystallites oriented differently with respect to the applied magnetic field.

Fluctuating local moments from surrounding magnetic ions will also cause inhomogeneous broadening. In the paramagnetic phase, they are fluctuating with the frequency $1/\tau_c$, where $\tau_c$ is the correlation time of the fluctuations. If the magnetic ions are fluctuating with a much higher frequency than the NMR linewidth, each nucleus will only sense the time-averaged local field and the NMR line is narrow (*motional narrowing*). When the magnetic correlations gain in strength and fluctuations slow down, the nuclei will begin to perceive a distribution of local fields. This will lead to a broadening of the NMR (and NQR) resonance lines (see Eq. (2.63) and Fig. 2.10 for the mathematical expression of the relation between $\tau_c$ and the linewidth $\Delta\omega \propto 1/T_2$).

# 3 NMR in the Superconducting State

This Chapter lays the foundations for the interpretation of the NMR measurements in the superconducting state which will be presented in Chapter 7. The behavior of the spin shift for spin-singlet and spin-triplet superconductors will be discussed in Section 3.1. Possible other shift contributions will also be pointed out. Section 3.2 starts with a detailed deduction of the spin-lattice relaxation rate of a BCS superconductor in the superconducting state. After deducing the spin-lattice relaxation rate of a single-band BCS superconductor and comparing it to the ultrasonic attenuation, some additional remarks about other possible gap symmetries will be shortly made. As will be pointed out, the comparison between NMR and ultrasonic attenuation was a significant proof of the BCS theory. At the end of Section 3.2 it will be shown what one should expect for the spin-lattice relaxation rate in a two-band model for $s_{++}$ and $s_{\pm}$ symmetries of the superconducting order parameter, since these symmetries will be important for the discussion of the spin-lattice relaxation rate measurements on pnictides later on.

The normal-state behavior of a superconductor will not be discussed theoretically in a separate chapter, but directly when discussing the experiments in Chapter 6.

## 3.1 Knight Shift

### 3.1.1 Spin Shift

Measuring the spin susceptibility in the superconducting state is an important tool to determine the pairing state of the Cooper pairs. An antiparallel alignment of the spins of the two electrons forming a Cooper pair (singlet pairing) leads to a total spin of $S = 0$. The spin susceptibility thus decreases in the superconducting state. A parallel orientation of the two spins yields a total spin of $S = 1$ (triplet pairing) which does not result in a change in the total spin susceptibility, at least not along the direction of the aligned Cooper spins. Since bulk magnetization measurements are dominated by the diamagnetic response of the superconductor, the spin part of the Knight shift is a good method to measure the spin susceptibility locally.

The spin shift is proportional to the static spin susceptibility, mediated via the hyperfine coupling, and can be expressed as [87]:

$$K_{s,sc} \propto \chi_{s,sc} = -4\mu_B^2 \int_0^\infty N_{sc}(E) \frac{\partial f(E)}{\partial E} dE \,, \tag{3.1}$$

where $N_{sc}$ is the density of states in the superconducting state and $f(E)$ is the Fermi distribution function. By comparing the spin susceptibility in the superconducting state with the Pauli spin susceptibility in the normal state, one obtains the ratio between the spin shift in the superconducting and in the normal state as:

$$\frac{K_{s,sc}}{K_{s,n}} = \frac{\chi_{s,sc}}{\chi_{s,n}} = -2\int_0^\infty \frac{N_{sc}(E)}{N_n(E_f)}\left(\frac{\partial f(E)}{\partial E}\right) dE\,. \tag{3.2}$$

**In the case of a BCS superconductor** with a single isotropic superconducting energy gap $\Delta$, the density of states in the superconducting state $N_{sc} = N_{BCS}$ is given by [2]:

$$\frac{N_{BCS}(E)}{N_n(E_f)} = \begin{cases} \frac{E}{\sqrt{E^2-\Delta^2}} & (E > \Delta) \\ 0 & (E < \Delta) \end{cases}. \tag{3.3}$$

The behavior of the spin shift in the superconducting state of a BCS superconductor can be described by the Yosida function derived from Equations (3.2) and (3.3) [88]. It describes the decrease of the spin shift with decreasing temperature below $T_c$. At low temperatures the decrease is exponential. For $T \to 0$ the spin shift vanishes: $K_s \to 0$ [88], expressing the gapping out of quasiparticle states.

**In the case of spin-triplet superconductors**, the spin susceptibility remains finite for certain directions. The order parameter in this case is characterized by the $\vec{d}$-vector. It contains information about the orientation of the $S = 1$ Cooper pairs and about the magnitude of the superconducting energy gap. The direction of $\vec{d}$ defines a normal to a plane to which the spins of the Cooper pairs are confined. Usually, spin-orbit coupling pins the $\vec{d}$-vector to a crystalline axis. Depending on the relative orientation of the external magnetic field $H_0$ versus $\vec{d}$, the spin shift either vanishes or remains finite [78]. Even though it is in practice a more complicated system, $Sr_2RuO_4$ is a famous example. Its $\vec{d}$-vector is believed to be: $\vec{d}(\vec{k}) = \Delta_0(k_x + ik_y)\hat{z}$, confining the spins of the Cooper pairs into the $xy$-plane. In this case $\chi_s$ and therewith $K_s$ remain constant across $T_c$ for $H_0 \parallel x$ and $H_0 \parallel y$ [21]. For $H_0 \parallel z$, a full Yosida function behavior is expected for $K_s$ [78, 89]. However, NMR measurements at very low fields (due to the very low upper critical field $H_{c2}$ along $z$) showed also a constant Knight shift along $z$ [90]. A re-orientation of the $\vec{d}$-vector (and thus of the spin direction) in small magnetic fields was discussed as a possible explanation for this unexpected behavior [90].

### 3.1.2 Orbital Shift and Spin-Orbit Scattering

Experimentally it is often observed that the measured Knight shift of singlet superconductors does not vanish for $T \to 0$, as expected theoretically, but decreases to some constant offset [91]. This effect stems from the orbital contribution to the Knight shift, $K_{orb}$, which is proportional to the orbital van Vleck susceptibility. It does not depend on the spin state of the electrons and is temperature-independent. While $K_s$ decreases exponentially to zero in the superconducting state for a singlet superconductor, the orbital shift $K_{orb}$ stays constant and gives rise to a finite NMR shift $K$ as $T \to 0$. In the case of strong orbital magnetism, $K_{orb}$ can become larger than the spin shift $K_s$ and lead to an unchanged Knight shift across $T_c$ even for singlet superconductors [92, 93]. Another effect to take into account is spin-orbit scattering, which mixes spin-up and spin-down states. In the presence of disorder, strong spin-orbit coupling might also cause a finite Knight shift in singlet superconductors [94].

Careful examination of the orbital contribution is thus needed before associating the observation of a constant Knight shift in the superconducting state straightforwardly to a triplet pairing symmetry.

### 3.1.3 Diamagnetic Shielding

In the superconducting state, the total shift $K_{\text{total}}$ is extended by a third contribution, stemming from the diamagnetic shielding currents induced by the external magnetic field $H_0$ [95, 96]:

$$K_{\text{total}} = K_s(T) + K_{orb} - \frac{\Delta H(T)}{H_0}. \tag{3.4}$$

The demagnetization field $\Delta H$ is a measure of the reduction of the internal magnetic field due to screening currents. The decrease of the resonance frequency due to the reduction of the internal magnetic field is expressed in the negative sign in front of the diamagnetic shielding contribution. This contribution is temperature- and field-dependent. It always leads to an additional decrease of the shift. If for instance the hyperfine field $A_{hf}$ of a certain nuclei would be negative, $K_s$ of a singlet superconductor would increase below $T_c$ [see Eq. (2.21)]. A diamagnetic shift of the same magnitude could then lead to the observation of a constant total shift and thus hide the actual decrease of the spin susceptibility. Experimentally, the quantitative contribution of diamagnetic shifts to the total shift is difficult to estimate. It can be quite large, which might handicap the correct interpretation of the Knight shift data in the superconducting state. An estimation of the importance of the demagnetization effect can be deduced from the calculation of the demagnetization factors [97]. Experimentally the observation of a line broadening due to the inhomogeneous field distribution in the mixed state evidences a large diamagnetic shielding effect. To eliminate the diamagnetic shielding effect and determine the real spin shift $K_s$, one has to measure the Knight shift at at least two different nuclei and subtract the different Knight shifts from each other. This is however only possible if both chosen nuclei measure the same spin degree of freedom [96].

## 3.2 Spin-Lattice Relaxation Rate

### 3.2.1 BCS Coherence Factors

The following deduction of the spin-lattice relaxation rate in the superconducting state of BCS superconductors is mainly based on a textbook by M. Tinkham [98], as well as on lecture notes by H. Eschrig [99] and J. Wosnitza [100].

In a BCS superconductor, the BCS ground state wave function is given as:

$$|\Psi_{BCS}\rangle = \prod_{\vec{k}} (u_{\vec{k}} + v_{\vec{k}} c^+_{\vec{k}\uparrow} c^+_{-\vec{k}\downarrow}) |\Phi_0\rangle , \tag{3.5}$$

where $|\Phi_0\rangle$ denotes the vacuum state without particles, $c^+_{\vec{k}\uparrow}$ and $c^+_{-\vec{k}\downarrow}$ are the creation operators of the electron states $(\vec{k}\uparrow)$ and $(-\vec{k}\downarrow)$ and $|v_{\vec{k}}|^2$ and $|u_{\vec{k}}|^2 = 1 - |v_{\vec{k}}|^2$ are the probabilities of finding an occupied and unoccupied Cooper pair $(\vec{k}\uparrow, -\vec{k}\downarrow)$, respectively. A general perturbation Hamiltonian can be written as:

$$\mathcal{H}_1 = \sum_{\vec{k}\sigma,\vec{k}'\sigma'} B_{\vec{k}'\sigma',\vec{k}\sigma} c^+_{\vec{k}'\sigma'} c_{\vec{k}\sigma} . \tag{3.6}$$

In the normal state, each pair of creation and annihilation operators $c^+_{\vec{k}'\sigma'} c_{\vec{k}\sigma}$ is independent from the others and transition probabilities between the one-electron states $(\vec{k}'\sigma')$ and $(\vec{k}\sigma)$ are proportional to the square of the matrix elements $B_{\vec{k}'\sigma',\vec{k}\sigma}$. On the other hand, in the superconducting state one has to deal with a phase-coherent superposition of one-electron states, which leads to interference terms in the perturbing Hamiltonian of the form of:

$$B_{\vec{k}'\sigma',\vec{k}\sigma} \left( c^+_{\vec{k}'\sigma'} c_{\vec{k}\sigma} \pm c^+_{\vec{-k}-\sigma,} c_{-\vec{k}'-\sigma'} \right) . \tag{3.7}$$

Whether the two terms will be summed up or subtracted depends on the nature of the particular perturbation Hamiltonian $\mathcal{H}_1$, which determines whether the two corresponding matrix elements have the same sign or not: $B_{\vec{k}'\sigma',\vec{k}\sigma} = \pm B_{\vec{-k}-\sigma,-\vec{k}'-\sigma'}$. Expressing the operators of Eq. (3.7) in terms of the fermionic quasiparticle operators above the BCS condensate (*Bogoliubons*), which are defined by the Bogoliubov-Valatin transformation[1] [99, 101, 102]:

$$\begin{aligned} \hat{b}_{\vec{k}\sigma} &= u_{\vec{k}} \hat{c}_{\vec{k}\sigma} - \sigma v_{\vec{k}} \hat{c}^+_{\vec{-k}-\sigma} \\ \hat{b}^+_{\vec{k}\sigma} &= u_{\vec{k}} \hat{c}^+_{\vec{k}\sigma} - \sigma v_{\vec{k}} \hat{c}_{\vec{-k}-\sigma} , \end{aligned} \tag{3.8}$$

and correspondingly [99]:

$$\begin{aligned} \hat{c}_{\vec{k}\sigma} &= u_{\vec{k}} \hat{b}_{\vec{k}\sigma} + \sigma v_{\vec{k}} \hat{b}^+_{\vec{-k}-\sigma} \\ \hat{c}^+_{\vec{k}\sigma} &= u_{\vec{k}} \hat{b}^+_{\vec{k}\sigma} + \sigma v_{\vec{k}} \hat{b}_{\vec{-k}-\sigma} , \end{aligned} \tag{3.9}$$

one ends up in coherence factors of the form of:

$$(uu' \mp vv')^2 = \frac{1}{2}\left(1 \mp \frac{\Delta^2}{EE'}\right) \tag{3.10}$$

$$(vu' \pm uv')^2 = \frac{1}{2}\left(1 \pm \frac{\Delta^2}{EE'}\right) . \tag{3.11}$$

Eq. (3.10) refers to a scattering of quasiparticles, while Eq. (3.11) refers to the annihilation or creation of two quasiparticles. For the expression of these coherence factors in terms of $E$ and $\Delta$, the mathematical trick:

$$\begin{aligned} u_k &= \sin\theta_k \\ v_k &= \cos\theta_k \\ \sin 2\theta_k &= \frac{\Delta_k}{E_k} \end{aligned} \tag{3.12}$$

[1] This transformation contains the constraint, that if one wants to describe only one excited quasiparticle $(\vec{k}\sigma)$ out of a superconducting Cooper pair $(\vec{k}\sigma, -\vec{k}-\sigma)$, one has to annihilate simultaneously its partner $(-\vec{k}-\sigma)$, since the excitation of one quasiparticle always entails the excitation of its Cooper pair partner.

was used, which is conveniently introduced to minimize the BCS Hamiltonian under the constraint $|u_{\vec{k}}|^2+|v_{\vec{k}}|^2 = 1$ [98]. $E_k = \sqrt{(\Delta_k^2+\xi_k^2)}$ is the excitation energy of a quasiparticle, $\Delta_k = \Delta$ is the isotropic energy gap and $\xi_k = \epsilon_k - \mu$ is the single particle energy relative to the Fermi energy.

Since in the case of quasiparticle annihilation/creation one quasiparticle obtains a negative energy relative to the chemical potential $\mu$, it is a common sign convention to reverse the sign in Eq. (3.11), such that the coherence factor for scattering as well as the one for quasiparticle annihilation and creation are given by:

$$C_{\mp} = \frac{1}{2}\left(1 \mp \frac{\Delta^2}{EE'}\right). \tag{3.13}$$

The sign depends on the nature of the interaction. If it is even in time reversal symmetry as in the case of electron-phonon interaction leading to ultrasonic absorption (expressed by a simple scalar potential), $C_-$ has to be used. $C_+$ is the right choice for interactions with odd time reversal symmetry, including vector potentials or spin-flip processes [98, 103].

### 3.2.2 BCS Spin-Lattice Relaxation Rate

The energies of spin flip scattering processes leading to spin-lattice relaxation are small compared to the superconducting energy gap, $\hbar\omega_L \ll \Delta$. Therefore, the spin-lattice relaxation is determined by quasiparticle scattering and not by quasiparticle creation/annihilation. The corresponding perturbing interaction involving spin-flip processes is odd under time reversal symmetry such that the spin-lattice relaxation rate is given by:

$$\alpha_s \propto |B|^2 \int_{-\infty}^{\infty} C_+(E,E')N_{BCS}(E)f(E)N_{BCS}(E')(1-f(E'))dE \tag{3.14}$$

Here, $f(E)$ is the Fermi distribution function, $N_{BCS}(E)$ the BCS density of states in the superconducting state [see Eq. (3.3)] and $C_+(E,E')$ is the formerly deduced coherence factor [see Eq. (3.13)]. $|B|^2$ is the transition matrix element containing the coupling of the nuclear spins to the electronic system (hyperfine coupling).

Experimentally, one can only measure the net transition rate between the energy levels $E' = E + \hbar\omega$ and $E$, which is given by the difference of the transition rate from $E$ to $E'$, $\alpha_{s,E\to E'}$ and the one back from $E'$ to $E$, $\alpha_{s,E'\to E}$:

$$\begin{aligned}\alpha_{s,E\leftrightarrow E'} &= \alpha_{s,E\to E'} - \alpha_{s,E'\to E} \\ &\propto N_{BCS}(E)f(E)N_{BCS}(E')[1-f(E')] \\ &\quad - N_{BCS}(E')f(E')N_{BCS}(E)[1-f(E)] \\ &= N_{BCS}(E)N_{BCS}(E')[f(E)-f(E')]\,. \end{aligned}\tag{3.15}$$

Furthermore, only processes with an energy difference $E' - E = \hbar\omega$ are interesting. Inserting the definition of the coherence factor $C_+$ from Eq. (3.13) into Eq. (3.14) and considering Eq. (3.15), the spin-lattice relaxation rate reads:

$$\alpha_s \propto |B|^2 \int_{-\infty}^{\infty} \frac{E(E+\hbar\omega)+\Delta^2}{E(E+\hbar\omega)} N_{BCS}(E)N_{BCS}(E+\hbar\omega)[f(E)-f(E+\hbar\omega)]dE\,. \tag{3.16}$$

In the normal state, the spin-lattice relaxation rate is given by:

$$\alpha_n \propto |B|^2 N_n^2(E_f)\hbar\omega\,. \tag{3.17}$$

For the interpretation of the spin-lattice relaxation rate in the superconducting state, it is sufficient to observe the changes which occur when the system enters the superconducting state. It is thus enough to determine the ratio $\alpha_s/\alpha_n$. Considering the ratio of the density of states given in Eq. (3.3), one arrives at:

$$\frac{\alpha_s}{\alpha_n} = \frac{2}{\hbar\omega}\int_\Delta^\infty \frac{E(E+\hbar\omega)+\Delta^2}{\sqrt{E^2-\Delta^2}\sqrt{[(E+\hbar\omega)^2-\Delta^2]}}[f(E)-f(E+\hbar\omega)]dE\,. \tag{3.18}$$

The factor 2 accounts for the integration from $-\infty$ to $-\Delta$ which contributes the same amount to the integral as the integration from $\Delta$ to $\infty$. Since NMR frequencies lie in the range of MHz and thus in the low energy range, one can approximate $\hbar\omega \ll \Delta$ and $\hbar\omega \ll k_BT$. Therewith $E+\hbar\omega \approx E$ and Eq. (3.18) simplifies to:

$$\begin{aligned}\frac{\alpha_s}{\alpha_n} &= \lim_{\hbar\omega\to 0}\frac{2}{\hbar\omega}\int_\Delta^\infty \frac{E(E+\hbar\omega)+\Delta^2}{\sqrt{E^2-\Delta^2}\sqrt{[(E+\hbar\omega)^2-\Delta^2]}}[f(E)-f(E+\hbar\omega)]dE\\ \frac{\alpha_s}{\alpha_n} &= 2\int_\Delta^\infty \frac{E^2+\Delta^2}{E^2-\Delta^2}\left(-\frac{\partial f}{\partial E}\right)dE\,.\end{aligned} \tag{3.19}$$

Calculating this integral by considering the temperature dependence of $f(E)$ results in the temperature dependence of the spin-lattice relaxation rate. Even without solving the integral completely, some qualitative temperature dependences can directly be deduced. Eq. (3.19) shows a logarithmic divergence from the integration at $\Delta$. This results in an increase of the spin-lattice relaxation rate just below $T_c$, the so-called Hebel-Slichter peak [18, 19]. At low temperatures, the spin-lattice relaxation rate follows an exponential temperature dependence $\alpha_s \propto \exp(-\Delta/k_BT)$, resulting mainly from the exponential tails of the function $-\partial f/\partial E$ [19].

The logarithmic divergence expressed in Eq. (3.19), which was calculated for $\omega \to 0$ overestimates the height of the Hebel-Slichter peak. If one considers a finite $\omega$ the divergence is smoothed by a factor of the order of $\ln(\Delta/\hbar\omega) \approx 10$, which is still too much compared to the experimentally observed rise in $T_1^{-1}$ of the order of 2. This can be explained by focussing on the dependence on the density of states. As seen already in Eq. (3.18), the spin-lattice relaxation rate is proportional to the square of the density of states:

$$\alpha \propto \int N(E)N(E+\hbar\omega)dE \approx \int N^2(E)dE\,. \tag{3.20}$$

The superconducting density of states $N_{BCS}(E)$ shows a divergence at $\Delta$ [see Eq. (3.3)]. The overestimation of the sharpness of this divergence by using the simple BCS expression (3.3) leads to an overestimation of the singularity in $\alpha_s/\alpha_n$. To account for a real system, anisotropies in the energy gap due to imperfections in the crystal structure as well as finite lifetimes of the quasiparticles (uncertainty principle) have to be considered, both causing a broadening of the singularity peaks in the density of states and thus a reduction of their height, which in the end reduces the height of the theoretical singularity in $\alpha_s/\alpha_n$ to the experimentally observed Hebel-Slichter peak. Practically, this can be done by convoluting the density of states with a breadth function $\sigma(E)$. The easiest breadth function is a rectangular one of width $2\delta E$ and height $1/2\delta E$ and was already proposed by Hebel and Slichter in 1959 [19].

Alternatively, one can rewrite Eq. (3.19) in the form:

$$\frac{\alpha_s}{\alpha_n} = \frac{2}{k_B T} \int_{\Delta}^{\infty} \frac{N_{BCS}^2(E) + M_{BCS}^2(E)}{N_n^2(E_f)} \left( -\frac{\partial f}{\partial E} \right) dE \,, \tag{3.21}$$

where $N_{BCS}(E)$ is the superconducting density of states as described above and $M_{BCS}(E)$ is the so-called anomalous density of quasiparticle states resulting out of the coherence factor [95, 104]. Instead of convoluting these densities of states with a rectangular breadth function, a broadening of the energy levels can also be introduced into the imaginary part of the density of states which then will be described as follows [105]:

$$N_{BCS}(E) = \mathrm{Re} \left\{ \frac{(E - i\Gamma)}{[(E - i\Gamma)^2 - \Delta^2]^{1/2}} \right\} \tag{3.22}$$

$$M_{BCS}(E) = \mathrm{Re} \left\{ \frac{\Delta}{[(E - i\Gamma)^2 - \Delta^2]^{1/2}} \right\} . \tag{3.23}$$

The anisotropy of the gap due to anisotropies in the phonon spectrum and/or the band structure can be expressed as [87]:

$$\Delta(\Omega) = \Delta_0[1 + a(\Omega)] \,, \tag{3.24}$$

where the anisotropy function $a(\Omega)$ with $\langle a(\Omega) \rangle = 0$ describes the variation of the gap at a certain direction $\Omega$ of the Fermi surface.

#### 3.2.2.1 Comparison to Ultrasonic Attenuation

The counterpart of the spin-lattice relaxation rate concerning the involvement of the coherence factor are measurements of ultrasonic attenuation. In this case, the coherence factor $C_-$ has to be used. Following the same deduction as in the case of the spin-lattice relaxation rate, but inserting $C_-$ instead of $C_+$, the ratio between the attenuation in the superconducting and in the normal state reads:

$$\begin{aligned} \frac{\alpha_s}{\alpha_n} &= \lim_{\hbar\omega \to 0} \frac{1}{\hbar\omega} \int_{\Delta}^{\infty} [f(E) - f(E + \hbar\omega)] dE \\ &= - \int_{\Delta}^{\infty} \frac{\partial f(E)}{\partial E} dE \\ &= \frac{2}{1 + \exp(\Delta/k_B T)} . \end{aligned} \tag{3.25}$$

Together with the temperature dependence of the superconducting energy gap $\Delta(T)$, this results in a drop of $\alpha_s/\alpha_n$ with infinite slope just below $T_c$ and in an exponential decrease for $T \ll T_c$. This was first observed and related to the BCS theory by Morse and Bohm in 1957 [106].

The Hebel-Slichter coherence peak in the spin-lattice relaxation rate together with the totally different behavior of the ultrasonic attenuation in the superconducting state played a significant role in establishing the BCS theory as the first microscopic description of conventional (weakly coupled) superconductors. The results, which are not describable simultaneously by any one-electron theory, are a proof of essential features of the BCS theory such as the existence of Cooper pairs and the superconducting gap, as well as of the actual BCS parameters $u_{\vec{k}}$ and $v_{\vec{k}}$ which are the basic ingredients of the coherence factors on the microscopic level.

### 3.2.3 Absence of the Hebel-Slichter Peak

**For a *d*-wave symmetry superconductor**, no coherence factors have to be taken into account. Correspondingly $M_{BCS}(E)$ vanishes [95]. Furthermore, the density of states in the superconducting state differs from the one of a BCS superconductor. Eq. (3.21) changes to:

$$\frac{\alpha_s}{\alpha_n} = \frac{2}{k_B T} \int_{\Delta}^{\infty} \frac{N_{d-sc}^2(E)}{N_n^2(E_f)} \left( -\frac{\partial f}{\partial E} \right) dE \,, \tag{3.26}$$

where $N_{d-sc}(E)$ is the density of states in the superconducting state of a *d*-wave superconductor. In a *d*-wave symmetry, the superconducting energy gap is $\vec{k}$-dependent and there are nodes on the Fermi surface, where it vanishes completely. This reduces the singularity in the density of states at $E = \Delta$ compared to $N_{BCS}(E)$ and additionally leads to a linear E-dependence of $N_{d-sc}(E)$ at low energies with $N_{d-sc}(E) \to 0$ for $E \to 0$. Altogether, this results in the absence of the Hebel-Slichter coherence peak and in a non-exponential decrease of the spin-lattice relaxation rate at low temperature [95, 107]. In a typical *d*-wave superconductor $T_1^{-1}$ decreases proportionally to $T^3$ for $T$ sufficiently below $T_c$. Deviations from the $T^3$-dependence may arise for $T \ll T_c$ due to pair breaking effects by impurities which produce a residual density of states at the lowest energies. Typically, this results in a linear temperature dependence of the spin-lattice relaxation rate at low temperature. Also crystal imperfections, vacancies, lattice distortions and similar defects may lead to a linear temperature dependence of $T_1^{-1}$ at low temperature.

However one has to be very careful not to associate straightforwardly the absence of the coherence peak with a *d*-wave symmetry. The peak might also be suppressed in a conventional *s*-wave superconductor. The occurrence of the Hebel-Slichter peak refers to single-band BCS superconductors in a *weak* coupling regime. In the *strong* coupling limit[2], where the electron-phonon coupling is enhanced, the Hebel-Slichter peak can be suppressed. An example of the effect of the coupling regime are the Chevrel-phase superconductors $TlMo_6Se_{7.5}$ ($T_c = 12.2\,K$) and $Sn_{1.1}Mo_6Se_{7.5}$ ($T_c = 4.2\,K$) [108]. While the latter one shows the Hebel-Slichter peak and an exponential decay of $T_1^{-1}$ with $2\Delta = 3.6 k_B T_c$, pointing towards an *s*-wave symmetry in the *weak* coupling limit[3], $TlMo_6Se_{7.5}$ does not exhibit any Hebel-Slichter peak, but its spin-lattice relaxation rate also decays exponentially with $2\Delta = 4.5 k_B T_c$ at low temperature. The value of 4.5 points towards a *strong* coupling regime, where the Hebel-Slichter coherence peak is suppressed by the lifetime effect of the quasiparticles through electron-phonon interactions, while the isotropic opening of the gap (*s*-wave symmetry) induces the exponential decay of $T_1^{-1}$.

[2] The general treatment of the effect of coupling strength is done by applying the Eliashberg theory, which is the generalization of the BCS theory to arbitrary coupling strengths.

[3] A classical BCS gap fulfills the relation $2\Delta = 3.52 k_B T_c$.

Another famous example is superconducting $A_3C_{60}$ (A = alkali metal). For this class of superconductors the Hebel-Slichter peak is suppressed by the application of high magnetic fields, suggesting an *intermediate* coupling regime within the BCS theory [96]. Other possible effects which could lead to a suppression of the Hebel-Slichter peak are a spread in superconducting transition temperatures within a sample or slight gap anisotropies [96].

While the existence of a Hebel-Slichter peak points strongly towards a conventional BCS $s$-wave pairing mechanism, the absence of it is not sufficient to interpret spin-lattice relaxation rates correctly in terms of other gap symmetries. Additional information is gained from the temperature dependence of the spin-lattice relaxation rate below $T_c$.

### 3.2.4 What to Expect for Pnictides?

A consideration of the $s_\pm$- and $s_{++}$-symmetry cases can be done in a two-band model, which is the most simple model which one can use to describe the pnictides. The coherence factors in this case are more complicated due to the two-band situation [103]:

$$C^{(i)} \propto \frac{1}{2}\left\{(\cdots)(\cdots) - \frac{\Delta_\nu \Delta_{\nu'}}{E_{\nu',\vec{k}} E_{\nu,\vec{k}+\vec{q}}}\right\} \tag{3.27}$$

$$C^{(ii)} \propto \frac{1}{2}\left\{(\cdots)(\cdots) + \frac{\Delta_\nu \Delta_{\nu'}}{E_{\nu',\vec{k}} E_{\nu,\vec{k}+\vec{q}}}\right\} . \tag{3.28}$$

Here, the focus is on the contribution of the superconducting gaps $\Delta_\nu$ and $\Delta_{\nu'}$ of the two bands $\nu$ and $\nu'$ (see [103] for the full expression of the coherence factors). Again, the sign depends on what one wants to look at (density response function or spin response function) according to the time reversal symmetry of the external field as a decisive parameter. The sign chosen in Equations (3.27) and (3.28) refers to the spin response function, which is the important one for NMR and neutron scattering. For the density response function, the sign has to be inverted in both equations. Furthermore, the sign depends on which regime one wants to focus on [103]: *(i)* $T \ll T_c$ and $\omega \approx |\Delta_\nu^{(0)}| + |\Delta_{\nu'}^{(0)}|$, where quasiparticle creation and annihilation are the leading terms [Eq. (3.27)]; and *(ii)* $0 \leqslant T \leqslant T_c$ and $\omega \ll |\Delta_\nu^{(0)}| + |\Delta_{\nu'}^{(0)}|$, where mainly the scattering of thermally excited quasiparticles is important [Eq. (3.28)] (see [103] for details). The interesting region for NMR measurements is always *(ii)*, since the Larmor energy lies in the range of $\hbar\omega \approx \mu$eV, while superconducting gap energies always are in the range of meV[4] and thus $\omega \ll |\Delta_\nu^{(0)}| + |\Delta_{\nu'}^{(0)}|$ is always true.

Most importantly, the sign in Equations (3.27) and (3.28) depends on the symmetry of the superconducting gap. This allows a differentiation between the $s_\pm$- and the $s_{++}$-symmetry by examining the product $\Delta_\nu \Delta_{\nu'}$. In the case of $s_\pm$-symmetry, $\Delta_\nu \Delta_{\nu'}$ is negative for $\vec{q} = \vec{Q}$, where $\vec{Q}$ is the nesting vector between the hole and the electron Fermi surface.

[4] Some examples of superconducting energy gaps:
bulk aluminum: $\Delta = 0.16\,\text{meV}$ [109],
$Bi_2Sr_2CaCu_2O_8$: $\Delta = 24\,\text{meV}$ [110],
LiFeAs: $\Delta_1 = 1.5 - 2.5\,\text{meV}$ and $\Delta_2 = 2 - 3.5\,\text{meV}$ [111],
$Ba_{1-x}K_xFe_2As_2$: $\Delta_1 = 9.1\,\text{meV}$ and $\Delta_2 = 1.5\,\text{meV}$ [112].

This leads to an effective sign reversion in front of the term $\Delta_\nu \Delta_{\nu'} / E_{\nu',\vec{k}} E_{\nu,\vec{k}+\vec{q}}$ at $\vec{q} = \vec{Q}$, which is responsible for the resonance peak in inelastic neutron scattering experiments [113–116], which focusses on the $\omega$-dependence of the spin response function around $\omega \approx |\Delta_\nu^{(0)}| + |\Delta_{\nu'}^{(0)}|$ and thus refers to Eq. (3.27). In the case of $s_{++}$-symmetry, $\Delta_\nu \Delta_{\nu'} > 0$ and no resonance peak would appear in inelastic neutron scattering experiments, at least considering only the coherence effects. However, Onari and coworkers showed later that a resonance-peak-like feature in inelastic neutron scattering experiments at $\omega \approx |\Delta_{min}^{(0)}| + |\Delta_{max}^{(0)}|$ can also be reproduced within an $s_{++}$ symmetry, when quasiparticle damping effects are included [117].

In NMR experiments, for which Eq. (3.28) has to be applied, the absence of the Hebel-Slichter coherence peak in the spin-lattice relaxation rate cannot be explained within the $s_{++}$ scenario, since the coherence factor $C^{(ii)}$ contains a positive sign as in the simple single-band BCS case. This sign will not be inverted in the $s_{++}$ symmetry case, since $\Delta_\nu \Delta_{\nu'} > 0$ in this case. Furthermore, the density of states in the $s_{++}$ case is very robust against impurities [103], in contrast to the $s_\pm$ scenario. Therefore, even by taking into account a smearing out of the singularity in the density of states due to impurities, the absence of the Hebel-Slichter coherence peak cannot be explained within the $s_{++}$ symmetry .

In the $s_\pm$ symmetry, in contrast, the absence of the Hebel-Slichter coherence peak follows directly out of the sign inversion in $C^{(ii)}$ due to $\Delta_\nu \Delta_{\nu'} < 0$.

Again it should be noted that the absence of the Hebel-Slichter coherence peak solely is not enough to prove a certain gap symmetry. An analysis of the temperature dependence of the spin-lattice relaxation rate is also needed, as well as the interpretation of NMR data in the context of other symmetry sensitive measurements, such as angle resolved photo emission spectroscopy (ARPES), neutron scattering, scanning tunnelling spectroscopy or specific heat.

# 4 Iron-based Superconductors

This Chapter gives a short overview of the recently discovered family of iron-based superconductors and a comparison to the cuprates, the first and most profoundly studied family of high-temperature superconductors. Additionally, a short overview of theoretical calculations of the electronic band structure, expected magnetic instabilities and the Fermi surface topology will be given. A knowledge of the latter is necessary for the understanding of possible gap symmetries and the impact of intraband and interband impurity scattering processes, mainly when discussing the temperature dependence of the spin-lattice relaxation rate in the superconducting state of $LaO_{0.9}F_{0.1}FeAs_{1-\delta}$ in Section 7.2.

Some basic properties of the studied materials $LaO_{1-x}F_xFeAs$ and LiFeAs measured by techniques other than NMR will be mentioned, as to give a proper context to the NMR measurements that will be presented later on. Particular properties of the specific samples which were measured in the course of this thesis will be presented in Chapter 5 or in direct comparison to the NMR results in Chapters 6 and 7.

## 4.1 General Overview

### 4.1.1 Crystal Structure and Electronic Structure

Iron-based superconductors can be divided into several structural families. The four most representative ones are depicted in Fig. 4.1. They are frequently named according to the stoichiometries of their parent compounds as "1111", "122", "111" and "11" systems. The common structural feature of all of these families is the two-dimensional (2D) iron-pnictogen layer (or in the case of the "11" a 2D iron-chalcogen layer). These layers, which are where superconductivity will take place, can be intercalated by some rare earth-oxygen layer ("1111"), by some alkali ions ("111") or alkaline earth ions ("122").

The two-dimensionality of these materials is at first glance very similar to the one of the cuprates, which are also layered systems. Physical properties are therefore expected to have a highly two-dimensional character, too. There is however also a difference in structure between cuprates and iron-based superconductors. The $CuO_2$ layers in cuprates are nearly flat (apart from some buckling). The iron-pnictogen layers in Fe-based superconductors are much more three-dimensional (see Fig. 4.1). Pnictogen ions are arranged above and below the iron-plane in a tetrahedral symmetry. This special arrangement allows the iron atoms to be located closer to each other, in contrast to the copper atoms in the cuprates.

Also the electronic structure differs between cuprates and pnictides. In pnictides (formally: $Fe^{2+}$: $3d^6$), all five Fe $3d$ orbitals cross the Fermi level and add charge carriers [113, 118–124] [see Fig. 4.2(a)], while in cuprates ($Cu^{2+}$: $3d^9$) mainly one Cu $3d$ orbital ($d_{x^2-y^2}$) is important [125–127]. Pnictides feature a high density of states at the Fermi

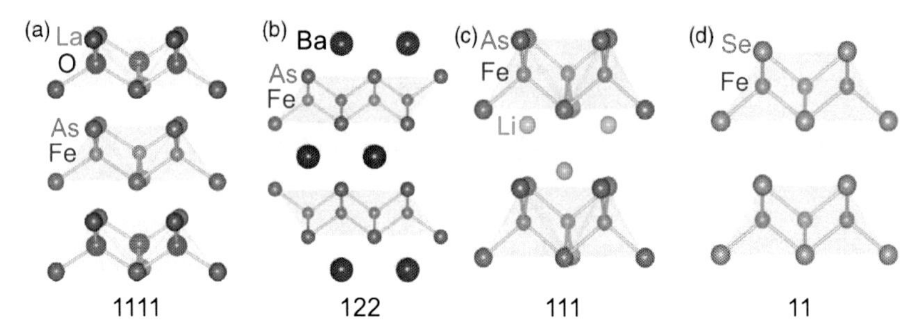


**Figure 4.1:** Crystal structure of the four main structural families of iron-based superconductors. Atom names correspond to representatives of the respective family. The iron-pnictogen layer (red-orange) is intercalated *(a)* by an RE-oxygen layer (blue-green) in the case of the "1111" family (e.g. LaOFeAs), *(b)* by an alkaline earth element (violet) in the case of the "122" (e.g. $BaFe_2As_2$), *(c)* by an alkali element (light green) in the "111" family (e.g. LiFeAs) or *(d)* by nothing in the case of the so-called "11" (e.g. FeSe). In this case, the pnictogen atoms are replaced by chalcogens to preserve the charge balance (graphic adapted from [10]).

energy, mainly dominated by Fe $3d$ bands [see Fig. 4.2(b)] whereas cuprates exhibit a rather moderate density of states at the Fermi energy [113, 118–124]. Furthermore, the hybridization of the pnictogen $p$ orbitals with the Fe $3d$ orbitals in pnictides results in a complicated electronic band structure and a multi-sheet Fermi surface composed of several small electron and hole Fermi surfaces which are disconnected from each other [see Fig. 4.2(c) for the Fermi surface topology in the folded[1] Brillouin zone], while in cuprates one observes only one large Fermi surface (at least at optimal doping) [113, 118, 119, 121–124]. Last but not least, the localized character of the Cu $3d^9$ holes and a very strong on-site repulsive Coulomb energy $U$ between different holes make the non-super-conducting parent compounds of cuprate superconductors become charge transfer insulators, while, thanks to their high density of states at the Fermi level, the non-superconducting parent compounds of pnictides are itinerant (semi)metals [123] and less correlated.

## 4.1.2 Ground States and Phase Diagrams

Most of the undoped parent compounds of iron-based superconductors (mainly the ones of the "1111" and "122" families) undergo a magnetic phase transition to an antiferromagnetically-ordered spin density wave (SDW) state at around $140 - 200\,\mathrm{K}$, which is accompanied or preceded by a structural phase transition from a tetragonal symmetry to an orthorhombic one. Upon hole- or electron-doping or upon applying pressure, these transitions get suppressed and superconductivity emerges. A notable exception is the "111" compound LiFeAs, which is a stoichiometric superconductor [14, 128, 129].

[1] The folded Brillouin zone corresponds to the crystallographic unit cell containing two Fe ions (and two As ions), to account for the fact that As ions are located above and below the FeAs-plane (see Fig. 4.1). The wave vector $(\pi, \pi)$ in the folded Brillouin zone corresponds to $(\pi, 0)$ in the unfolded Brillouin zone with one Fe atom per unit cell [113].

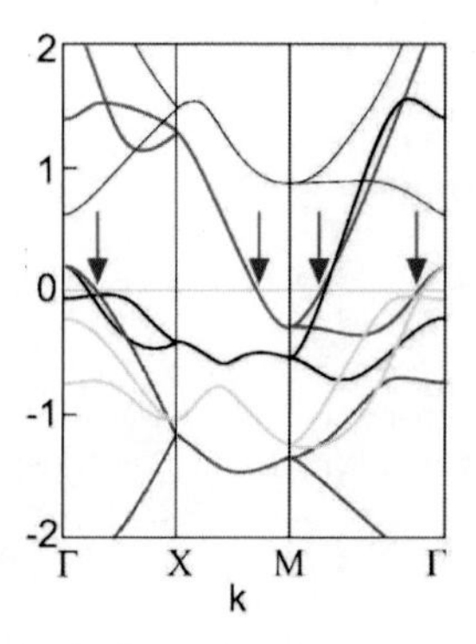


(a) LDA (local density approximation) band-structure calculations for LaOFeAs. Arrows denote the points where the bands cross the Fermi level. (Calculations from [119], figure reproduced from [113].)

(b) LDA calculations of total (black) and partial (colored) electronic density of states (DOS) of LaOFeAs. The DOS at the Fermi level is dominated by Fe $3d$ contributions. (Figure reproduced from [119].)

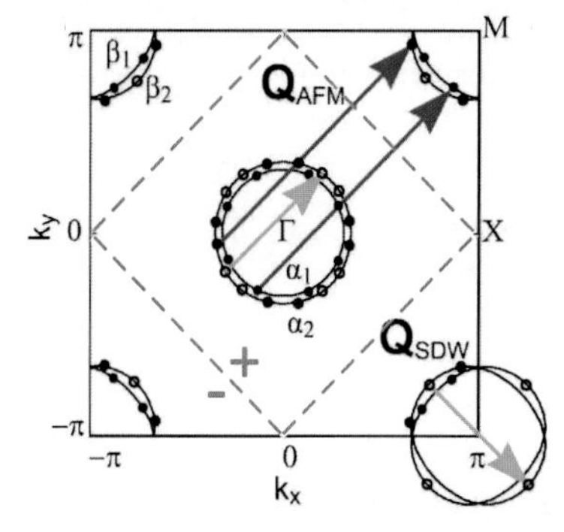


(c) Calculated Fermi-surface topology of LaOFeAs based on an effective four-band model. The Fermi surface consists of two hole pockets around $\Gamma = (0,0)$ and two electron pockets around $M = (\pi, \pi)$ in the folded Brillouin zone. Blue (green) arrows denote main scattering vectors with $\vec{Q}_{\mathrm{afm}} = (\pi, \pi)$ (interband) and with the incommensurate intraband scattering $\vec{Q}_{\mathrm{SDW}}$. The dashed light orange line and the signs depict the nodes and the corresponding sign of the superconducting $s_{\pm}$ order parameter (see Section 4.1.3 for the introduction of the symmetry of the superconducting order parameter). (Figure adapted from [113].)

**Figure 4.2:** Selected calculations of the electronic structure of pnictides for the example of LaOFeAs, highlighting the multiband character ($a$), the high density of states at the Fermi level dominated by Fe $3d$ bands ($b$) and the multi-sheet Fermi-surface topology ($c$). Similar figures can be found in the references listed in the text.

Electron-doping in the "1111" family is for instance achieved by substituting oxygen ($O^{2-}$) by fluorine ($F^{1-}$), leading to (RE)$O_{1-x}F_x$FeAs [5], by introducing oxygen deficiencies: (RE)$O_{1-\delta}$FeAs [130] or by substituting cobalt for iron: (RE)O$Fe_{1-x}Co_x$As [131]. In the "122" family, electron-doping is also achieved by substituting Fe partially with Co leading to A$(Fe_{1-x}Co_x)_2As_2$ with A = Ba, Ca, Sr [132]. Hole-doping in the "122" families is done by partially substituting the divalent alkaline earth element (A = Ba, Sr, Ca) with a monovalent alkali element (B = K, Cs, Na), leading to $A_{1-x}B_xFe_2As_2$ with the famous example of $Ba_{1-x}K_xFe_2As_2$ [12].

The parent compounds of cuprate superconductors also exhibit a transition to an antiferromagnetic ground state (at somewhat higher temperatures), but in contrast to the SDW state in pnictides, the magnetically-ordered state in cuprates is a localized charge transfer insulator. Very similar to pnictides, superconductivity emerges upon doping cuprates with positive or negative charge carriers.

Concerning the doping-dependence of the magnetic and the superconducting phase transitions, it is possible to describe all cuprate superconductors by one universal phase diagram, where the transition temperatures are plotted versus the concentration of electron (left side of the phase diagram) and hole dopants (right side of the phase diagram) [see Fig. 4.3(a)] [126, 133, 134]. In the case of iron-based superconductors, no unified phase diagram could be established up to now. The properties of each material vary with the choice of the interlayer. This is mostly pronounced within the "1111" family, where, depending on the rare earth (RE) ion, different phase diagrams have been reported [see Fig. 4.3(b) and 4.3(c)]. The question of coexistence or separation of the magnetic and superconducting phase in the crossover region between these two phases is still under debate [135–139]. Also the coexistence of the superconducting phase with the static magnetism of some rare earth ions at low temperature and the interplay between the RE and the Fe magnetism are highly studied topics [140–143].

### 4.1.3 Symmetry of the Superconducting Order Parameter

The Cooper-pairing mechanism in high-temperature superconductors in general is one of the most challenging problems in contemporary solid state physics. The search for the symmetry of the superconducting order parameter in pnictides is an additional challenge.

Historically, NMR studies that revealed the Hebel-Slichter peak, caused by the coherence factor and the symmetry of a single nodeless superconducting gap, played a significant role in establishing the BCS theory as the first microscopic description of conventional (weakly-coupled) superconductors [19] (see Chapter 3). Nowadays, within a simplified approach (ignoring damping, strong coupling, anisotropy, impurity, and inhomogeneity effects [144, 145]) its presence or absence together with the temperature dependence of the spin-lattice relaxation rate below $T_c$ are frequently used to discriminate tentatively conventional from unconventional pairing. For a single Fermi surface sheet and superconductivity in the clean limit $T^3$- and $T^5$-dependencies would be regarded as evidence for line- and point-node superconducting order parameters, respectively [146], which for singlet pairing correspond to the $d$- and a special $s + g$-wave state.

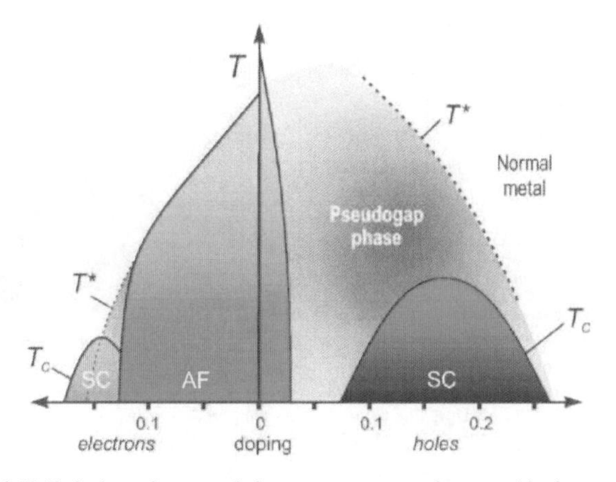


(a) Unified phase diagram of the cuprate superconductors with electron doping on the left side and hole doping on the right side, showing the superconducting (SC) and antiferromagnetic (AF) phase as well as the pseudogap phase. (Figure reproduced from [134].)

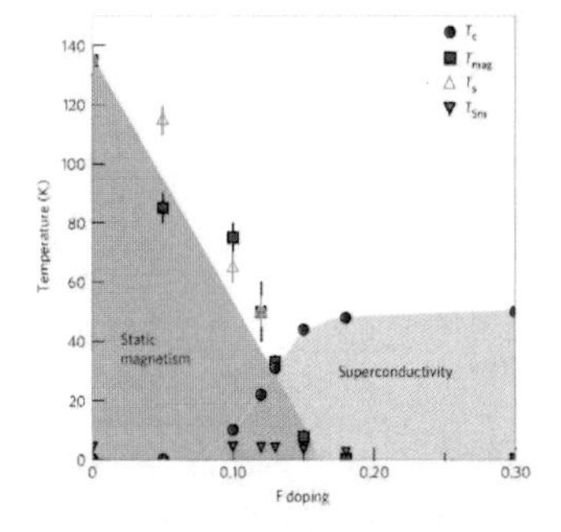


(b) Phase diagram of $SmO_{1-x}F_xFeAs$, comprising the structural ($T_s$), magnetic ($T_{mag}$) and superconducting ($T_c$) phase transition as well as the ordering temperature of the Sm moments ($T_{Sm}$). (Figure reproduced from [137].)

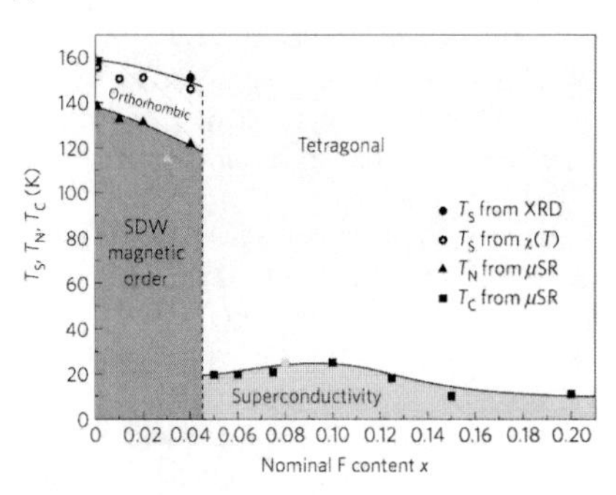


(c) Phase diagram of $LaO_{1-x}F_xFeAs$, with structural ($T_s$), magnetic ($T_N$) and superconducting ($T_c$) transition temperatures. (Figure reproduced from [135].)

**Figure 4.3:** Comparison of the unified phase diagram for cuprate superconductors (*a*) with two representative phase diagrams of electron-doped "1111" materials, namely $SmO_{1-x}F_xFeAs$ (*b*) and $LaO_{1-x}F_xFeAs$ (*c*), which cannot be generalized to a common phase diagram of the "1111" family. While in $SmO_{1-x}F_xFeAs$ a clear region of coexistence of magnetism and superconductivity is observed, magnetism and superconductivity are strictly separated in the case of $LaO_{1-x}F_xFeAs$.

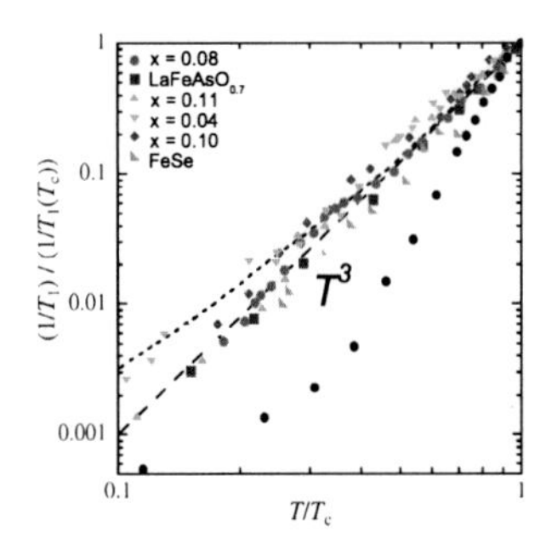


**Figure 4.4:** Frequently observed $T^3$-dependence of $T_1^{-1}$ in the superconducting state of different samples of $LaO_{1-x}F_xFeAs$, $LaO_{0.7}FeAs$, and FeSe. (Fig. adapted from [162].)

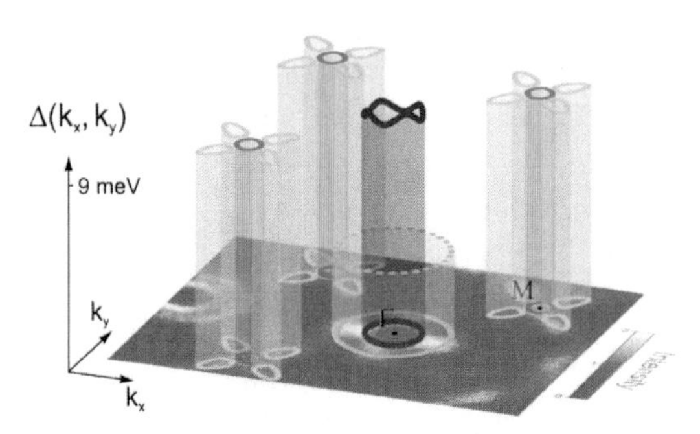


**Figure 4.5:** Momentum dependence of the superconducting gaps on the different Fermi surfaces in $Ba_{1-x}K_xFe_2As_2$ with underlying Fermi surface intensity map as measured by ARPES. The height denotes the magnitude of the gaps. (Fig. reproduced from [163].)

Therefore, early NMR studies on Fe-based superconductors related the generally observed $T^3$ dependence of the spin-lattice relaxation rate in the superconducting state and the absence of a Hebel-Slichter peak (see Fig. 4.4) to an unconventional line-node pairing symmetry [79, 147–150].

These findings stood in contrast to the results of various other experiments that have been carried out to extract the symmetry of the superconducting order parameter in Fe-based superconductors. In particular, angle-resolved photo-emission spectroscopy (ARPES), Andreev reflection and microwave data [151–159] are consistent with a superconducting gap being nodeless on each Fermi surface pocket. These results taken together with the observation of a peak at the antiferromagnetic wave vector $\vec{Q}$ and $\omega = \omega_{res}$ found below $T_c$ in various compounds [116, 160, 161] by means of inelastic neutron scattering (INS) experiments provide support in favor of $s_\pm$-wave symmetry.[2]

The $s_\pm$ wave-symmetry, proposed by *Mazin et al.* [121], is a consequence of the multiple Fermi surface sheets in the electronic structure of the Fe-based superconductors, as plotted in Fig. 4.2(c). Within this model, the superconducting gaps which open at each Fermi surface upon entering the superconducting state are isotropic and similar in magnitude, but differ in their sign ($+\Delta$ and $-\Delta$ on the hole and the electron pocket, respectively) [121]. The pairing interaction is thus repulsive (but still leads to pairing due to the sign change).

Concerning NMR, soon it was realized that the situation in these multiband compounds, especially in the presence of impurities, is far from being that simple. In particular, there is no universal behavior for the growing number of related compounds. Power-law dependencies $T_1^{-1} \propto T^n$ with $n$ in between 1.5 and 6 have been observed in different pnictides in the course of time [79, 147–150, 162, 164–173]. The only exception found up to now is $LaO_{0.9}F_{0.1}NiAs$, where $T_1^{-1}$ shows indeed a Hebel-Slichter coherence

[2] As discussed in Section 3.2.4 a sharp resonance peak is a result of different signs of the superconducting gap for $\vec{k}$ and $\vec{k} + \vec{Q}$ points generic for the $s_\pm$-wave symmetry.

peak followed by an exponential decrease [174]. However, this compound differs from the other ones in other important aspects. It shows a rather low $T_c$ of maximal 4 K and the absence of antiferromagnetic fluctuations [174].

The power-law dependencies of $T_1^{-1}$ clearly indicate unconventional superconductivity. Besides the possibility of a simple line-node gap, the results have also been discussed within other models, such as the $s_\pm$ wave-symmetry [121, 175–177] and the frequent inclusion of two superconducting gaps with different amplitudes [147, 165, 170, 171]. The $s_\pm$ wave symmetry model is able to describe the NMR data when considering additional impurity scattering effects (see dotted line in Fig. 4.4) [121, 175–177] and thus can reconcile NMR and ARPES results. A detailed discussion of the $T_1^{-1}$ data in the superconducting state of $LaO_{1-x}F_xFeAs$ and the theoretically proposed $s_\pm$ symmetry of the superconducting gap function will be given in Section 7.1.

Fig. 4.5 shows the momentum dependence of the superconducting gaps in $Ba_{1-x}K_xFe_2As_2$ as measured by ARPES [163]. Besides a slight anisotropy of the gap on the inner $\Gamma$ barrel the gaps are isotropic. The gaps on the inner $\Gamma$ barrel and on the pockets and the blades around $M$ amount to $\approx 9$ meV. The gap on the outer $\Gamma$ barrel is less than 4 meV, supporting the two-gap scenario.

Concerning the resonance peak in INS, it has been argued recently [117] that a somewhat broader peak-like feature can be attributed to a self-energy renormalization of quasiparticles in $s_{++}$-wave (sign preserved) superconductors. A similar feature in Raman spectra has not been observed [178]. Hence, the assignment of the observed INS features and the interpretation of NMR spin-lattice relaxation rate data in the superconducting state with regard to a certain pairing symmetry of the superconducting order parameter are still controversial.

## 4.2 Basic properties of $LaO_{1-x}F_xFeAs$

$LaO_{1-x}F_xFeAs$ is the archetype of the "1111" family and the iron-based superconductors in general, since it was the first pnictide where superconductivity has been found [5]. The question of interaction between the rare earth ion magnetism and the magnetism of the FeAs layer is not important in this compound, since the lanthanum ions do not have any $4f$ moments and are therefore non-magnetic. The detailed phase diagram was first reported by Luetkens *et al.* [135] and is reproduced in Fig. 4.6. The parent compound LaOFeAs shows a structural phase transition from a tetragonal symmetry to an orthorhombic one at $T_s = 156$ K, followed by a magnetic phase transition to an antiferromagnetic SDW state at $T_N = 138$ K. These phase transitions are observable by various experimental techniques such as measurements of susceptibility, resistivity or specific heat, X-ray diffraction, thermal expansion, $\mu$SR, and $^{57}$Fe Mössbauer spectroscopy [135, 179–183]. The magnetic moment in the SDW state deduced from early $^{57}$Fe Mössbauer spectroscopy and neutron scattering experiments lies in between 0.25 and $0.35\,\mu_B$ [183–186], which is far below the theoretically calculated value of $0.9-2.6\,\mu_B$ using DFT (density functional theory) in both, local moment or itinerant approaches [119, 187–191]. More recent neutron scattering measurements reported a magnetic moment of $0.63\,\mu_B$ [192], which is still lower than the theoretically expected value, but agrees very well with the value reported by NMR measurements [193], which will be presented in Section 6.1.1. The experimentally-observed

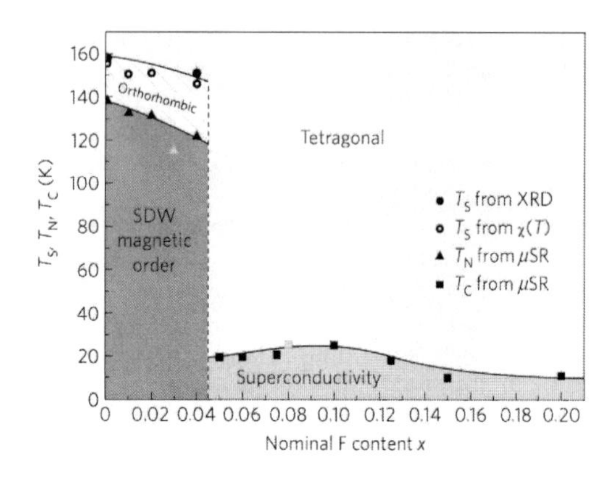


**Figure 4.6:** Phase diagram of $LaO_{1-x}F_xFeAs$ (reproduced from [135]).

reduced value of the magnetic moment is a common feature of the pnictides. In CeOFeAs, PrOFeAs and NdOFeAs, it is reported to range from 0.25 to 0.9 $\mu_B$ [136, 141, 194–198] while in $BaFe_2As_2$, $CaFe_2As_2$ and $SrFe_2As_2$ it lies in between 0.8 and 1 $\mu_B$ [198–203].

Upon doping LaOFeAs by substituting oxygen with fluorine (electron-doping), both $T_s$ and $T_N$ first decrease, before they vanish abruptly between 4% and 5% F-doping and superconductivity emerges for $x \geq 5\%$ [135]. The highest superconducting transition temperature is found for $LaO_{0.9}F_{0.1}FeAs$ ($T_c = 26$ K).

In bulk susceptibility measurements, the phase transitions are visible in pronounced anomalies at the corresponding transition temperatures $T_s$ and $T_N$ (see Fig. 4.7) [181]. In the normal state, the susceptibility decreases linearly with decreasing temperature with an intrinsic, doping-independent slope over the whole accessible doping-range, regardless of the nature of the underlying ground state [181]. A decreasing susceptibility with decreasing temperature seems to be a generic property of iron-pnictides in general. Several theoretical approaches including details of the Fermi surface, a temperature dependence of the density of states, pseudogap effects and others were discussed to explain such a decrease. A comparison between the bulk susceptibility and NMR shift data on $LaO_{0.95}F_{0.05}FeAs$ and $LaO_{0.9}F_{0.1}FeAs$ as well as a discussion of the possible origins of the observed decrease will be given in Section 6.1.

The calculated Fermi surface consists of four small, two-dimensional and disconnected Fermi surface sheets: two hole cylinders centered around the $\Gamma$ point (0,0) and two electron cylinders centered around $M = (\pi, \pi)$ [in the *folded* Brillouin zone, see Section 4.1.1 and Fig. 4.2(c)] [113, 118, 119, 121–123, 190]. Another additional 3D hole pocket derived from a hybridization between the Fe 3*d* orbitals and the As *p* and La orbitals, crosses the 2D hole pockets at $\Gamma$ [118, 123]. This pocket gets filled up quickly upon electron-doping and disappears just in the crossover region between the SDW state and the superconducting ground state in between $x = 0.04 - 0.05$ [121, 123]. Only the 2D hole and electron pockets are left, whose two-dimensionality puts constraints on the possible superconducting pairing symmetry. Strong variations of the order parameter along $k_z$ are very unlikely. Strong

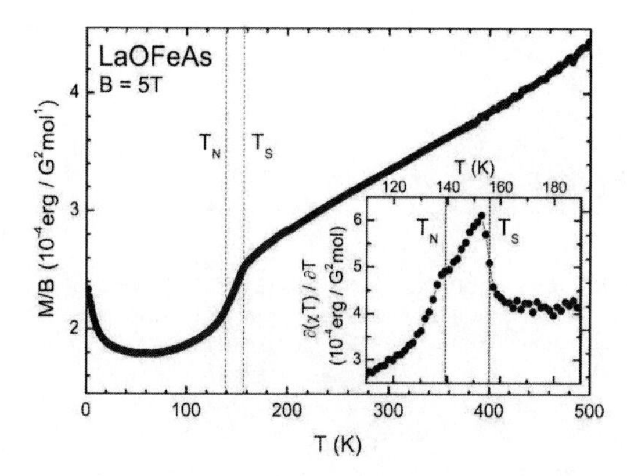


**Figure 4.7:** Temperature-dependent static susceptibility of undoped LaOFeAs, measured in a magnetic field of 5 T. The structural and magnetic phase transition at $T_s$ and $T_N$ lead to visible anomalies in the susceptibility itself and in its magnetic specific heat $\partial(\chi T)/\partial T$ (inset). (Figure reproduced from [181].)

variations of the order parameter on one Fermi surface sheet within the $xy$-plane are also unlikely, because they would require a strong $\vec{q}$-dependence of any possible pairing interaction, which is not easily compatible with the small size of each Fermi surface sheet [121]. In the undoped parent compound LaOFeAs the 2D electron- and hole-like Fermi surfaces match well if one of them is translated by the antiferromagnetic wavevector $\vec{Q} = (\pi, \pi)$ [see Fig. 4.2(c)]. Additionally, their orbital characters are similar. This good *nesting* would drive the system to the experimentally observed SDW instability [121–123, 190]. Upon fluorine doping, the nesting gets worse, leading to a suppression of the magnetic phase transition and apparently opening the field for superconductivity.

Angle-resolved photoemission spectroscopy (ARPES) studies on LaOFeAs report the theoretically expected high density of states at the Fermi level, the multiple band character and the existence of several hole and electron like Fermi surfaces [124, 204, 205]. However, there are some disagreements between the ARPES data and the existing band structure calculations. Due to charge redistribution effects in the "1111" family, several surface states contaminate the electronic structure measured by ARPES. Fig. 4.8 shows some of the ARPES results on LaOFeAs and the corresponding distinction between bulk electronic and surface electronic states [204].

The superconducting pairing mechanism is still unsettled. Conventional electron-phonon coupling was found to be too weak to account for the rather high superconducting transition temperature [120].

Slightly different gap values have been reported by different experimental methods until now, depending on the underlying symmetry assumptions and if one or two gaps have been used to fit the data. Table 4.1 summarizes the gap values reported by photoemission studies (PES) [206], Point-contact Andreev reflection measurements (Andreev) [157] and fits to the temperature-dependent $^{75}$As NMR and NQR $T_1^{-1}$ [165, 168]. All these data have been collected by measuring polycrystalline samples. If a two-gap model was used

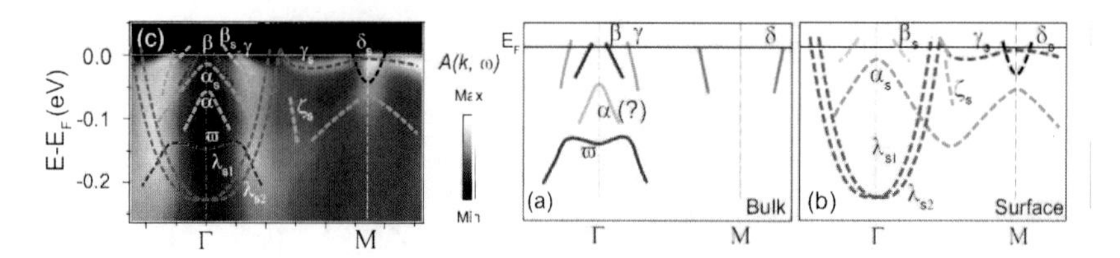


**Figure 4.8:** ARPES intensity plot along the direction Γ-M ($c$) vs the summary of the bulk ($a$) and the surface ($b$) band structures. Dashed curves in ($c$) are guides to the eyes obtained by tracking the local minimum of the second derivative of the raw data with respect to energy. Figure adopted from [204].

to extract the gap sizes, the second value in parentheses refers to the second gap. The last column of Table 4.1 gives the value $2\Delta/k_BT_c$. It is always close to the value 3.53 for a weakly-coupled BCS superconductor (at least for one gap).

| sample | method | gap value(s) (meV) | $2\Delta/k_BT_c$ |
|---|---|---|---|
| $LaO_{0.93}F_{0.07}FeAs$ | PES [206] | 3.6 ($s$-wave) | 3.5 |
| | | 4.1 ($d$-wave) | 3.96 |
| $LaO_{0.9}F_{0.1}FeAs$ | Andreev [157] | 4.6 | 3.44 |
| | | 4.4 (10-12) | 3.23 (8.5) |
| $LaO_{0.92}F_{0.08}FeAs$ | NQR $T_1^{-1}$ [165] | | 3.75 (1.5) ($s_{\pm}$-wave) |
| | | | 4.2 (1.5) ($d$-wave) |
| $LaO_{0.89}F_{0.11}FeAs$ | NMR $T_1^{-1}$ [168] | | 4 |

**Table 4.1:** Values for the superconducting gap in $LaO_{1-x}F_xFeAs$ as reported by photoemission spectroscopy (PES), point-contact Andreev reflection (Andreev) and $^{75}$As NQR and NMR $T_1^{-1}$. If two gaps were used to fit the data, the value of the second gap is put in parentheses. The underlying gap symmetry assumption is also put in parentheses.

## 4.3 Basic properties of LiFeAs

The physical properties of LiFeAs differ quite a bit from the ones of the "1111" and "122" families. The crystal structure is the same as for the "1111" family, namely $P4/nmm$. But in contrast to other pnictides, LiFeAs is a stoichiometric superconductor. No evidence for any structural or magnetic phase transition has been found. Superconductivity in single crystals of LiFeAs is reported to set in at $T_c = 18\,\mathrm{K}$ [14, 128, 129]. The superconducting ground state is of multiband character and exhibits two nearly completely isotropic gaps with $\Delta_1 \approx 1.5\,\mathrm{meV}$ for the hole-like Fermi surfaces and $\Delta_2 \approx 2.5\,\mathrm{meV}$ for the electron-like Fermi surface sheets [111, 207]. Compared to $T_c$, the largest gap is close to the BCS weak-coupling limit, $2\Delta_2/k_BT_c = 3.22$.

The absence of a magnetic phase transition and the direct onset of superconductivity in stoichiometric LiFeAs is surprising since DFT calculations find the electronic structure of LiFeAs to be very akin to those of other pnictides. They report a high density of states at the Fermi level which is dominated by the Fe 3*d* bands and a similar shape of the Fermi surface with hole and electron cylinders at the Brillouin zone center and corners, respectively [208–210]. The theoretically calculated ground state of stoichiometric LiFeAs is an antiferromagnetic SDW state very similar to that of the "1111" and "122" families [208–210]. Additionally, LiFeAs should be even further away from a ferromagnetic instability than LaOFeAs due to its slightly lower density of states at the Fermi energy [208]. Theoretical calculations of the phonon dispersion modes and the electron-phonon coupling strength found an electron-phonon coupling strength of $\lambda = 0.29$ [210]. Applying McMillan's formula [211, 212], which relates $\lambda$ to the superconducting transition temperature $T_c$ revealed that $\lambda = 0.29$ is too small to account for the experimentally observed $T_c$ of 18 K [210]. As in the case of other pnictide families, electron-phonon coupling alone is thus not enough to explain the relatively high $T_c$.

The discrepancy between the experimental absence of a magnetic ground state and the DFT predictions was discussed in terms of interlayer spin coupling [209] or possible deviations from perfect stoichiometry in the investigated samples [210].

ARPES measurements on LiFeAs are easier to perform than on LaOFeAs, since the cleavage plane lies in between two layers of Li atoms, so that the surfaces are not polarized. The observed Fermi surface map is very similar to the Fermi surface topology of other pnictides, consisting of five sheets, three hole-like ones around the Γ point (a large one and two small ones) and two electron-like ones at the Brillouin zone corners (see Fig. 4.9) [111]. However, in contrast to other pnictide parent compounds, nesting between these different Fermi surfaces is completely absent. Furthermore, a strong renormalization (factor 3) is needed to scale the experimental data with the calculated band structure [111]. The two small hole-like Fermi surfaces around Γ are formed by the Fe $3d_{xz}$ and $3d_{yz}$ bands. These are very flat bands showing nearly no dispersion near the Fermi energy, which leads to an enhanced density of states at the Fermi level (van Hove singularity) (see Fig. 4.9) [111]. The existence of a van Hove singularity at the center of the Brillouin zone (Γ-point) has been proven by quasiparticle interference in scanning tunneling microscopy (STM) experiments [213].

The absence of any nesting and the strong renormalization might explain the experimentally-observed absence of magnetic order, since they were not included in the DFT calculations. Recent band structure calculations including the absence of nesting and the shallow character of the two small hole pockets do not find an antiferromagnetic ground state, but dominant "almost ferromagnetic" incommensurate fluctuations near (0,0) and thus a proximity to a ferromagnetic instability which might be activated by electron doping [214].

The proximity to a ferromagnetic instability is strengthened in the case of Ni-doping in LiFeAs. Upon Ni-doping, superconductivity is suppressed and a transition to a ferromagnetic ground state is observed (at $T_C = 156$ K for 2.5% of Ni). Detailed studies of the stoichiometry revealed that Ni-doping in LiFeAs goes along with the presence of Li-deficiencies. These deficiencies might actually be the most important ingre-

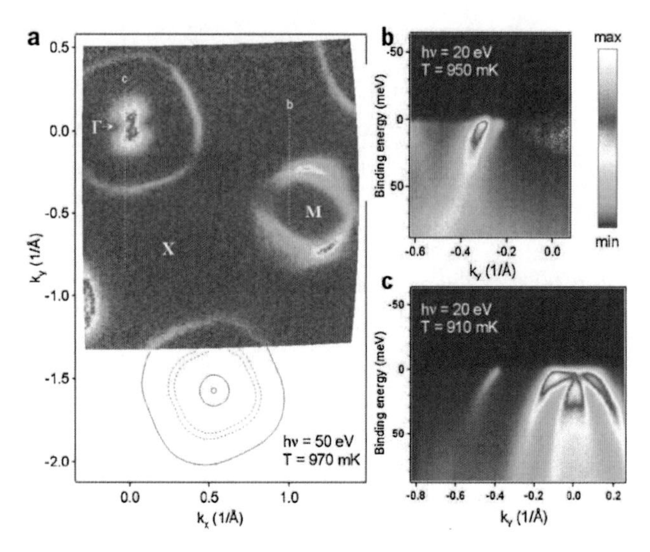


**Figure 4.9:** (*a*) Fermi surface map of LiFeAs as measured by ARPES, revealing three hole-like Fermi surface sheets around Γ and two electron-like around M. The sketch of these Fermi contours (Γ-centered contours as solid lines, M-centered contours as dashed lines) at the bottom shows the absence of nesting. (*b*) Momentum-energy cut along the direction b marked in (*a*). *c*) Momentum-energy cut along the direction c marked in (*a*). The tops of both small bands practically touch the Fermi level. The upper band shows nearly no dispersion, leading to an enhanced density of states at the Fermi level. Figure reproduced from [111].

dient for ferromagnetism, as revealed by the observation of a ferromagnetic transition at $T_C \approx 160\,\mathrm{K}$ in $Li_{1-x}Fe_{1+x}As$ with a nominal Li-deficit of 2%.[3] With increasing Ni content (Li-deficiencies), the ferromagnetic transition temperature decreases until the disappearance of ferromagnetism at 10% nominal Ni-doping. The magnetic moment in the ferromagnetic state deduced from susceptibility measurements amounts to $0.1\,\mu_B$ only, suggesting itinerant ferromagnetism.

Cobalt-doping in LiFeAs on the other hand seems to act as normal charge doping. It suppresses superconductivity (both $T_c$ and the superconducting volume fraction) in LiFeAs. Above a Co concentration of 5%, superconductivity disappears. No Li-deficiencies were found in the Co-doped samples [216].

Since the influence of Ni-doping and the effect of Li-deficiencies are still under examination, no phase diagram can be drawn for LiFeAs at the moment. The observations described in the two preceding paragraphs are based on unpublished, ongoing work at the IFW Dresden.

[3] The Fe excess might also play a role. Note however the high Curie temperature of single crystalline Fe: $T_C = 1043\,\mathrm{K}$ [215].

# 5 Experimental Setup

This Chapter will describe the main elements generally needed for solid state NMR/NQR experiments such as magnets, cryogenics, probes, and electronics and will specify which setup was used particularly for the measurements done in this work. An overview of the samples used for the measurements will be given. Various experimental problems had to be faced during the experiments. Some of them and their solutions will be presented in Section 5.4.

## 5.1 Magnets and Cryogenics

During the course of this thesis, all three available NMR magnets have been used. When starting in June 2008, only two warm bore magnets existed. One magnet (Bruker) has a permanent static field of $H = 7.0494\,\mathrm{T}$ and a homogeneity better than $\Delta H/H = 10^{-6}$. The other one (Magnex Scientific) can be fixed to any desired magnetic field value between 0 and 9.2 T by adjusting the current through its big superconducting coil. With another small superconducting coil, the field can then be continuously swept in a range of $\Delta H = \pm 0.2\,\mathrm{T}$. The homogeneity of this second magnet is 7 ppm over a 10 mm axial plot and the main field drift at 9.2 T is 0.545 ppm/h. Both warm bore magnets are equipped with continuous flow cryostats from Janis which enable measurements at temperatures ranging from 1.5 K to 325 K. For the high temperature measurements, a high temperature continuous flow cryostat from Oxford Instruments was used, which allowed measurements up to 500 K.

In August 2009 a new cold bore 16 T field sweep magnet was put into operation (Oxford Instruments). Its homogeneity over 10 mm diameter of a spherical volume is 11.8 ppm. In persistent mode, the current in the superconducting coil decays with 7 ppm/h.[1] Via a needle valve the variable temperature insert (VTI) of this magnet is directly connected to the helium bath which cools the superconducting coil. The temperature of the sample space is controlled by regulating the opening of the needle valve and - if needed - by pumping. Temperatures between 1.5 and 400 K can be reached with this system. The heater, the Cernox temperature sensor and the needle valve of the VTI are controlled with an Oxford ITC 502 temperature controller.

[1] The first superconducting coil which was delivered for this magnet had an homogeneity of 18 ppm and a current decay in persistent mode of 18 ppm/h. This coil was exchanged with the better one described in the text some months later.

## 5.2 Electronic Measurement Equipment

### 5.2.1 Probes

A resonant circuit for NMR/NQR measurements has to fulfill two conditions: it must be tunable to the desired frequency in the MHz-range and it must have an impedance of $50\,\Omega$ such that it couples optimally to the rest of the electronic measurement equipment. This ensures that the power is optimally coupled into and out of the resonant circuit, which leads to optimal sensitivity and avoids reflecting power back to the amplifier. Various sample probes were used for the NMR and NQR measurements, according to the desired frequency and temperature ranges. Each of these probes was equipped with one out of two different types of resonant circuits (see Fig. 5.1): one containing two variable glass capacitors for tuning and impedance matching [Fig. 5.1(a)]; and one containing a variable cylindrical capacitor for the tuning and a coil with a brass core for the matching [Fig. 5.1(b)]. The tuning to the measurement frequency is done by changing the capacity of the tuning-capacitor, while the impedance of $50\,\Omega$ is reached by changing the capacity (inductance) of the serial matching capacitor (parallel matching coil). A resonant circuit composed of glass capacitors is advantageous because it covers a very broad frequency range, due to their wide capacity range (2 to 120 pF in the case of our probes). The resonant circuit sketched in Fig. 5.1(b) is better suited for high power. But since the necessary power is rather small (see Section 5.4.3) mostly sample probes with glass capacitor resonant circuits were used. For instance, the measurements on the air-sensitive LiFeAs ranged from around 21 Mhz (NQR) up to 62 MHz (NMR). To minimize the contact with air, the whole frequency range had to be covered with one sample coil and one resonant circuit. This was only possible with a glass capacitor resonant circuit.

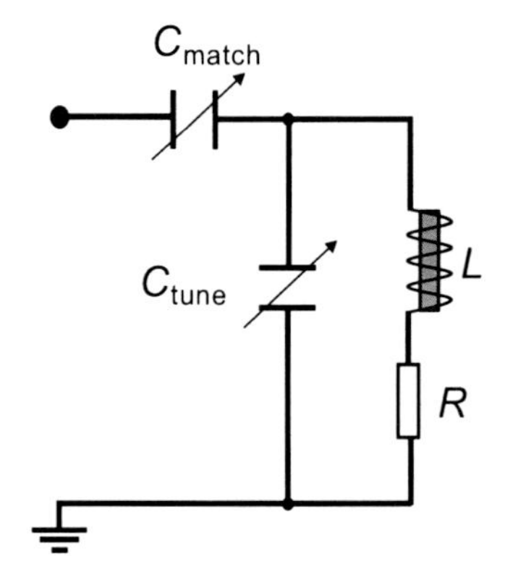

(a) glass capacitor resonant circuit

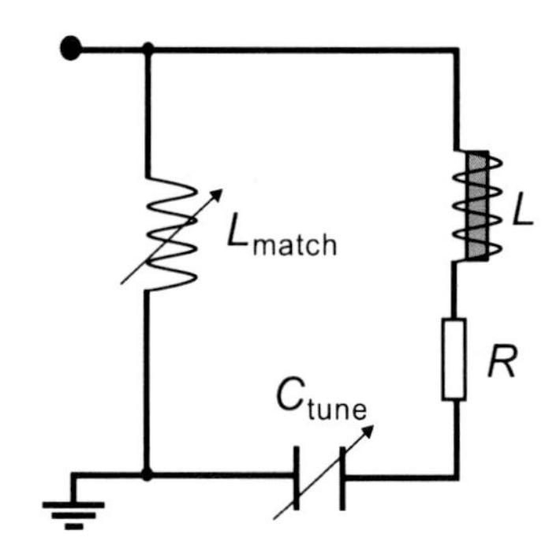


(b) resonant circuit with matching coil and cylindrical tuning capacitor

**Figure 5.1:** Both types of used resonant circuits. The sample (grey rectangular) is located inside the sample coil with inductance $L$ and residual serial resistance $R$.

## 5.2.2 Main Elements of the Electronics

Figure 5.2 combines the main electronic elements of a typical experimental NMR setup. Since high frequency (HF) pulses have to be generated and at the same time very weak echo signals from the sample have to be detected, the setup of the electronic measurement equipment is challenging. The spectrometer (Tecmag Apollo), which is connected to the computer via an USB-port and controlled by the program *NTNMR*, generates pulses of desired frequencies and shapes, executes pulse sequences and simultaneously detects and digitizes the signal from the sample. Its interior assembling is therefore rather sophisticated and will not be described in detail here. The pulses generated by the spectrometer are amplified to several hundred volts by the HF amplifier. A directional coupler separates a small part of the power coming from the HF amplifier (approximately 1%) and conducts this part to the oscilloscope, where the shape of the pulses and their reflections are monitored. Most of the pulse power is directed from the directional coupler to the transcoupler. The transcoupler separates the incoming high frequency pulses produced by the HF amplifier from the echo signal coming back from the sample, which amounts only to some $\mu$V. This separation is realized by a serial connection of two sets of crossed semiconductor diodes and a $\lambda/4$-cable in between them. The strongly nonlinear voltage-current behavior of the diodes let them act as switches between high and low voltage ranges, while the $\lambda/4$-cable produces an impedance transformation. Consequently, during the duration of the HF pulses, the transcoupler connects the HF amplifier to the sample probe, while directly after the HF-pulse application the transcoupler connects the sample probe to the preamplifier. The preamplifier and internal amplifiers of the spectrometer amplify the echo signal from the probe, before it is digitized and detected. In the case of the sweepable 16 T and partially sweepable 9.2 T magnets, their power supplies were connected to the computer via a serial port and controlled via a script of the *NTNMR* program. In some cases also the temperature controller (LakeShore) was connected to the computer to monitor the temperature and the heater output. Network analyzers (not shown in Fig. 5.2) were used before each measurement to tune and match the resonant circuit to the designated frequency.

The constituent hardware components are specified in the following.

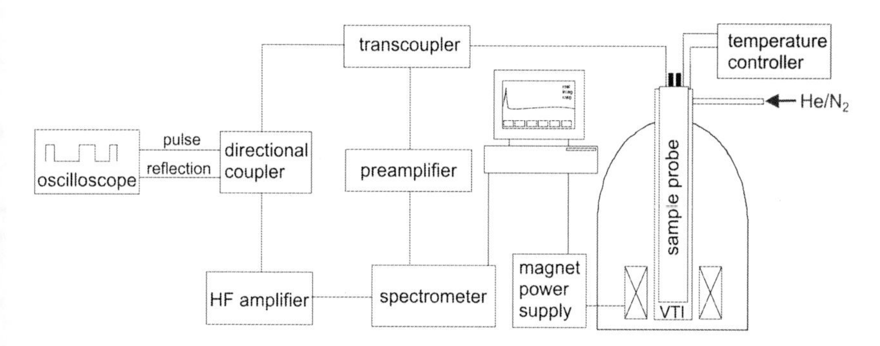


**Figure 5.2:** Schematic diagram of a typical experimental NMR setup (warm bore magnet).

In the case of the three NMR setups, Tecmag Apollo spectrometers were used in combination with the *NTNMR* program. In the beginning of this thesis, NQR measurements were done with one of these spectrometers, as time and demands permitted. At the end of the course of this work, the NQR setup got its own spectrometer, a LapNMR spectrometer from Tecmag, controlled by the program *NTNMR*.

Oscilloscopes from Agilent Technologies, Tektronix and Hameg were used for the pulse monitoring. Later on, an oscilloscope card was installed at the NQR measurement computer and the pulses were monitored by the program *GageScope*.

Network analyzers from Agilent Technologies and Hameg were used to tune the resonance circuits to the desired frequency. New hardware and the program *Advanced Stepper Control*, assembled and written by Yannick Utz, could be used for an automatic adjustment of the tuning and matching of the NQR sample probe. With another program (*AAS - Automatic Adjustment System*), also written by Yannic Utz, automated frequency sweeps could be realized at the NQR setup.

The temperature and the heaters were monitored and controlled with either LakeShore 331, LakeShore 340 or Oxford ITC 502 temperature controllers. The superconducting magnet power supplies were Oxford IPS 120-10 at the 16 T magnet and Cryomagnetics Cs-4 at the 9.2 T for the big coil (static field) and LakeShore 625 at the 9.2 T magnet for the small coil ($\pm 0.2$ T sweeps).

## 5.3 Samples

### 5.3.1 $LaO_{1-x}F_xFeAs$

The polycrystalline samples of $LaO_{1-x}F_xFeAs$ measured in the course of this thesis were grown at the IFW Dresden following and improving the two-step solid state reaction approach of Zhu *et al.* [179, 217]. The characterization by means of structural, thermodynamic and transport measurements was also done in house [179–181]. $^{57}$Fe Mössbauer spectroscopy and muon spin relaxation done by the group of Hans-Henning Klauss at the TU Dresden complemented the characterization [135, 183, 218]. Table 5.1 collects the structural, magnetic and superconducting transition temperatures $T_s$, $T_N$ and $T_c$ of the samples which were considered in this work [135, 179–181, 183, 218, 219].

| $x$ | $T_s$ | $T_N$ | $T_c$ |
|---|---|---|---|
| 0 | 156 K | 138 K | - |
| 0.05 | - | - | 20 K |
| 0.075 | - | - | 22 K |
| 0.1 | - | - | 26 K |
| 0.1 As-Def | - | - | 28.5 K |

**Table 5.1:** Transition temperatures $T_s$, $T_N$ and $T_c$ of $LaO_{1-x}F_xFeAs$ for all considered doping levels [135, 179–181, 183, 218, 219]. The last row (As-Def) corresponds to a special arsenic-deficient sample $LaO_{0.9}F_{0.1}FeAs_{1-\delta}$, whose enhanced superconducting properties will be discussed in Section 7.2.

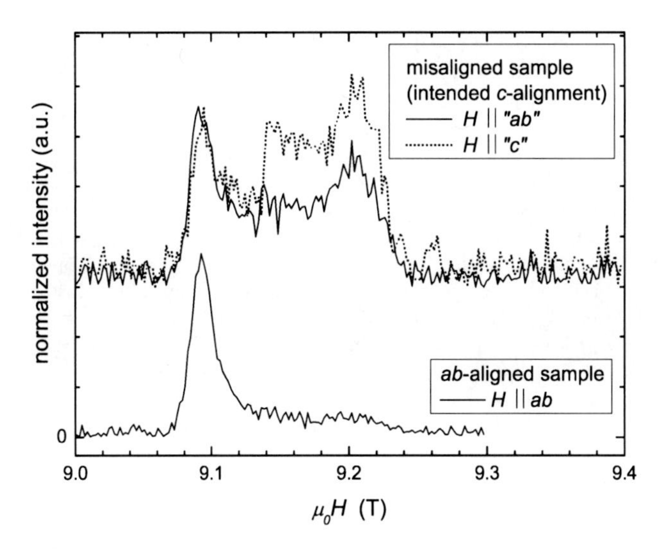


**Figure 5.3:** $^{75}$As field sweep spectra at room temperature for an aligned (lower part) and a misaligned (upper part) powder sample of $LaO_{0.9}F_{0.1}FeAs$.

For the NMR measurements, the pellets were ground to a powder with a rough grain size of 1 - 100 $\mu$m to maximize the surface area and therewith the signal-to-noise ratio in NMR-experiments[2] and to allow for orientation. Each powder was put in a quartz glass tube of 3 mm diameter, which was sealed with Teflon thread tape and wax or two-component adhesive to protect the samples against possible degradation by water and air. The mass of the samples varied between 50 and 120 mg. Some samples were also oriented in magnetic field to allow for directional measurements. For this purpose, the ground powders were mixed with Stycast epoxy 1266 in a mass ratio of 24:70. Directly after the mixing procedure this mixture was put into an external field of 9.2 T. Since the magnetic susceptibility of $LaO_{1-x}F_xFeAs$ is anisotropic, the randomly oriented grains of the powder align with their crystallographic axis having the highest susceptibility (which in this case is *ab*) along the direction of the external magnetic field. Letting the sample-epoxy mixture cure in the external field, well *ab*-oriented samples have been achieved. This procedure is very sensitive to the viscosity of the epoxy. An enhanced viscosity prevents the crystallites from aligning, while a reduced viscosity leads to a collection of the grains at the bottom, which also leads to misorientation. Some samples did not cure. In these cases the alignment was not satisfactory. An attempt to produce *c*-axis-aligned powder samples using the field-rotation alignment method described by Chang *et al.* [220] failed for $LaO_{0.9}F_{0.1}FeAs$. It worked for samples of $PrO_{1-x}F_xFeAs$ later on. The NMR measurements reported in the following chapters will therefore concentrate on powder

[2] The radiofrequency applied in NMR experiments is shielded by conducting samples. Grounding powders increases the surface area and such the total amount of accessible nuclear spins.

samples or *ab*-oriented samples of $LaO_{1-x}F_xFeAs$. Note that the *ab*-aligned crystallites within the oriented powder samples do not have a common *c*-axis. NMR measurements can therefore only be done along the orientation direction.

A good measure of the degree of orientation are central line spectra. Fig. 5.3 shows examples of well aligned and misaligned sample. The upper part of Fig. 5.3 shows the room temperature $^{75}$As field sweep spectra of an intended *c*-axis alignment for both $H \parallel ab$ and $H \parallel c$. Both spectra (taken for $H \parallel ab$ and $H \parallel c$, respectively) cover the same broad field range and resemble more or less normal powder spectra as reported in [79]. Only the relative enhancement of the $H \parallel c$ feature at around 9.14 T (see Appendix A.1 for a detailed discussion of powder spectra) for the $H \parallel c$ measurement (dotted line) gives a small indication of a partial *c*-axis-alignment, which is however not sufficient at all. The lower part of Fig. 5.3 shows the room temperature $^{75}$As field sweep spectrum of a well *ab*-aligned sample for $H \parallel ab$. It consists of one resonance line at the field value which corresponds to the $H \parallel ab$ peak of a powder spectrum. Only the slight intensity enhancement between 9.12 T and 9.24 T indicates that some minor parts of the sample have not been aligned. This sample can be used e.g. for Knight shift measurements, which concentrate on the well-defined maximum of the $H \parallel ab$ peak.

### 5.3.2 LiFeAs

Single crystals of LiFeAs were grown using the self-flux technique [129]. The stoichiometry has been checked with inductively coupled plasma mass spectroscopy and energy dispersive X-ray spectroscopy and found to be Li:Fe:As=0.99:1.00:1.00 [129]. Angle resolved photoemission spectroscopy agreed with an almost exact stoichiometry [111]. The high quality of the single crystals is reflected in all thermodynamic properties. The susceptibility shows a sharp superconducting transition at $T_c$ = 18 K. A complete diamagnetic shielding of the zero-field-cooled (ZFC) susceptibility reveals a 100 % superconducting volume fraction, thus bulk superconductivity [129]. Also the resistivity shows a very sharp superconducting phase transition of $\Delta T_c$ = 1.2 K. Even in high magnetic fields the transition width remains remarkably sharp. The residual resistivity is very low[3], $\rho_0$ = 15.2 $\mu\Omega$.cm, and the residual resistivity ratio (RRR) is larger than in all other pnictides [223]. The specific heat also shows a sharp superconducting transition and a negligibly small electronic contribution at low temperature [207].

Furthermore, the $^{75}$As NQR linewidth (FWHM) at room temperature is exceptional narrow. Fig. 5.4 shows the $^{75}$As NQR resonance line of a LiFeAs single crystal at room temperature. A Gaussian fit to the line yields a full width at half maximum (FWHM) of only $\Delta\nu$ = 44 kHz. This remarkably small linewidth[4] excludes the possibility of vacancies or interstitial sites and corroborates both the electronic and structural homogeneity of our samples. A polycrystalline sample of LiFeAs prepared in our institute shows an increased $^{75}$As NQR linewidth of about $\Delta\nu \approx$ 113 kHz at room temperature ($\Delta\nu \approx$ 130 kHz at

[3] For comparison: The residual resistivity of polycrystalline samples of LiFeAs lies in the range of some m$\Omega$cm [128]. For single crystals of Ni- or Co-doped $BaFe_2As_2$ the residual resistivity is $\rho_0 \geq 50\,\mu\Omega$.cm [221], and for Ni-doped $SrFe_2As_2$ single crystals it is $\rho_0 \geq 100\,\mu\Omega$.cm [222].

[4] For comparison: $\Delta\nu \approx$ 220 kHz for stoichiometric LaOFeAs [30, 193] and $\Delta\nu \approx$ 480 kHz for stoichiometric $CaFe_2As_2$ [224].

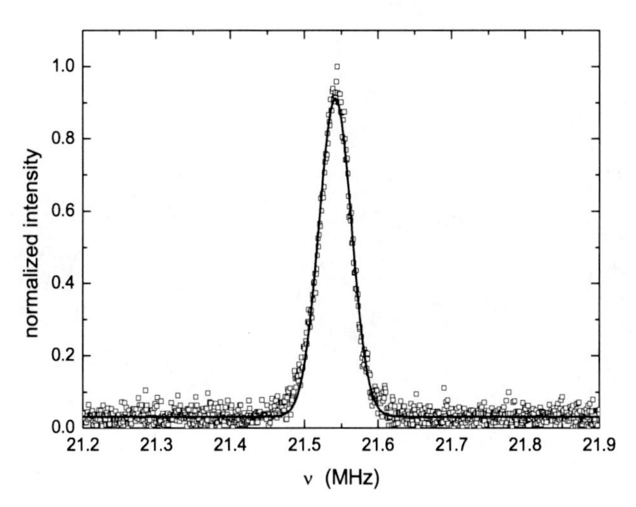


**Figure 5.4:** $^{75}$As NQR resonance line of a LiFeAs single crystal at room temperature (open squares). The solid line is a Gaussian fit to the data yielding a linewidth of $\Delta\nu = 44$ kHz.

$T = 20$ K) [129], while other groups reported even larger values of the $^{75}$As NQR linewidth of polycrystalline samples (e.g. $\Delta\nu \approx 170$ kHz at $T = 20$ K [81]).

The high quality of the pure LiFeAs single crystals is further confirmed by the $^7$Li-spectra (see Fig. 5.5). It is possible to resolve all three resonances of the $^7$Li spectra, although the quadrupole frequency deduced from these spectra is only $\nu_q = 32$ kHz and according to Fig. 2.2 the separation between the two satellites for $H \parallel ab$ ($H \parallel c$) is only $\nu_q$ ($2\nu_q$) (the principal axis of the EFG in LiFeAs is parallel to the $c$-axis: $V_{ZZ} \parallel c$). The linewidth is as small as 9 kHz (11 kHz) for $H \parallel ab$ ($H \parallel c$) and it is the same for the central transition and the satellites, excluding possible quadrupole effects on the lineshape due to disorder. A linewidth of 90 kHz at room temperature was reported on polycrystalline samples [80], where a resolution of the three-split $^7$Li-spectrum was not possible. An echo-decay measurement on these polycrystals yielded a quadrupole frequency of $\nu_q = 34$ kHz, which compares nicely to our results. Another group reported $^7$Li NMR measurements on LiFeAs single crystals [225]. They could also resolve the three peaks of the $^7$Li NMR spectrum and deduced a quadrupole frequency of about $\nu_q \sim 60$ kHz. However, they do not comment on their $^7$Li NMR linewidth and they observe a second, broad $^7$Li NMR resonance line at lower frequencies, possibly stemming from interstitial Li sites in their crystals. Such a second $^7$Li NMR resonance line has not been found in our crystals, underlining again the high crystal quality of our samples.

All LiFeAs crystals are highly air-sensitive and susceptible to exfoliation. They were kept inside a glovebox under argon atmosphere until the beginning of the measurements. The NMR sample coil was prepared in advance without the sample. It was made conform to the sample dimensions to ensure a high filling factor. The sample was put into the

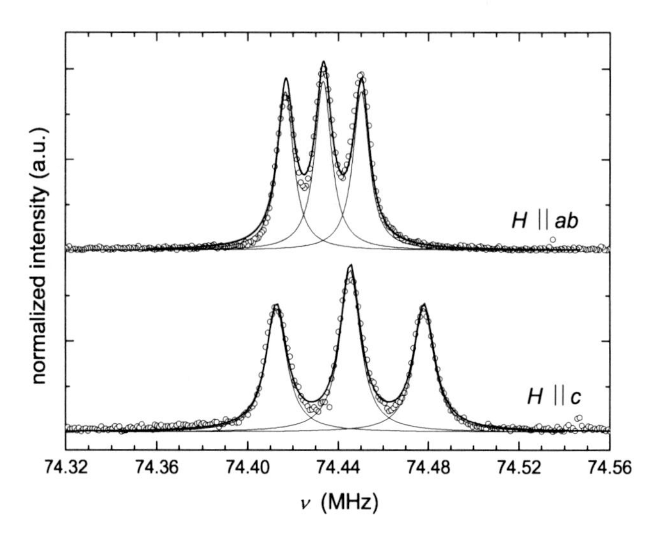


**Figure 5.5:** $^7$Li NMR spectra at 200 K measured in $H = 4.4994$ T for $H \parallel ab$ (upper part) and $H \parallel c$ (lower part). The solid lines are fits to three Lorentzian lines with the constraint that both satellites have the same linewidth, intensity and quadrupole shift $\nu_q$.

coil inside the glovebox. Both were then placed into a quartz glass tube and covered with Teflon thread tape. Once out of the glovebox the glass tube was immediately sealed with epoxy and the sample was mounted to the NMR sample probe. The probe was directly put into the cryostat, which always was under He atmosphere. Not all of the NMR measurements reported in this thesis were done at one and the same sample. Some samples degraded when the sample probe had to be taken out for a while (which was necessary to change capacitors for instance). The sample quality was therefore checked by $^{75}$As NQR measurements after each extraction of the sample probe. The crystal which was used for most of the NMR measurements had the dimensions $5.04 \times 2.24 \times 0.52$ mm. Other crystals had similar dimensions.

## 5.4 Problems and Improvements

### 5.4.1 Temperature Control

A well-defined control and monitoring of the sample temperature is one of the main requirements for correct NMR measurements. Because of the distance between the sample on one side and the heater and the temperature sensor at the bottom of the cryostat (VTI) on the other side, monitoring and regulating the temperature via the cryostat (VTI) was not exact enough.

This was particularly pronounced in the 16 T field sweep magnet, where the distance between the Cernox temperature sensor of the VTI and the center of the magnetic field,

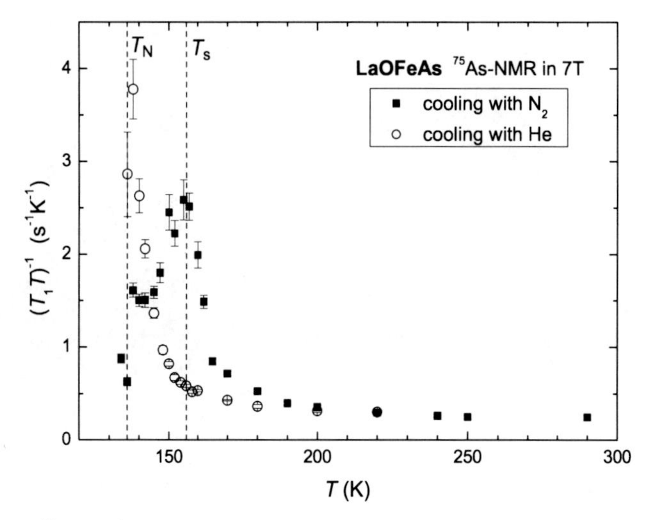


**Figure 5.6:** $^{75}(T_1T)^{-1}$ of LaOFeAs measured in $H = 7\,\mathrm{T}$, cooled with nitrogen (filled squares) and helium (open dots). The dashed vertical lines indicate the structural and magnetic transition temperature of LaOFeAs, $T_s$ and $T_N$, respectively.

where the sample is located, amounts to more than 4.7 cm. To account for this discrepancy, a second Cernox temperature sensor was installed directly in the sample probe close to the sample. Furthermore, the probe was equipped with a second heater in form of a wound resistance wire (with a resistance load of $23\,\Omega/\mathrm{m}$) around the whole cap of the sample probe [see Fig. 5.7(b)]. The resistance wire was doubled before winding, such that no additional magnetic fields due to loop currents were generated at the sample site. The total resistance of the new cap-heater was $86\,\Omega$. The rough temperature control of the VTI was henceforward done with the Oxford ITC 502, while the fine tuning of the temperature at the sample site was done with the LakeShore 331 reading the sensor near the sample and regulating the cap-heater of the sample probe. Since then, very stable temperature conditions were obtained. Another advantage of this additional temperature control and heating regulation is the reduced helium consumption, since the cap heating is rather localized. Not the whole VTI has to be heated, but only the region within the cap. Furthermore, heating the cap of the sample probe leads to very stable temperature conditions inside the cap, without large temperature gradients.

In the two warm bore magnets the temperature control was not such a big problem, since the distance between heater and sensor at the bottom of the continuous flow cryostat and the center of the magnetic field is smaller. The right temperatures were achieved successfully by cooling the sample space with helium[5]. Problems arose when nitrogen was used as a cooling medium Fig. 5.6 shows $^{75}$As NMR measurements on LaOFeAs. $(T_1T)^{-1}$

[5] This was checked e.g. by comparing superconducting or magnetic transition temperatures measured in the NMR magnets and measured in other setups.

exhibited a maximum at the structural transition temperature $T_s = 156\,\mathrm{K}$ instead of a divergence at the magnetic ordering temperature $T_N = 138\,\mathrm{K}$. Furthermore, a strange hysteresis between cooling and heating was observed. A check with helium as cooling medium displayed the expected divergence of $(T_1T)^{-1}$ towards $T_N$ and no hysteresis between cooling and heating (see Fig. 5.6). The reason for this strange observation could not be figured out. It is possible that the nitrogen flow was not optimally regulated. The temperature at the sensor of the bottom of the continuous flow cryostat was stable, though. A wrong regulation of the heating power can therefore be excluded as a source for the observed differences.

To overcome similar problems and ensure always the right sample temperature, irrespective of the use of helium or nitrogen as cryogenics, also the sample probes of the warm bore magnets were endowed with additional Cernox sensors close to the sample and resistance wire heaters around their caps in the course of time, resulting in very stable and easily controllable temperature conditions at the sample site.

### 5.4.2 Sample Probe for the 16 T Field Sweep Magnet

For the newly arrived 16 T field sweep magnet from Oxford Instruments, a new sample probe was designed in house. In the beginning of the measurements with this new sample probe, a lot of problems had to be faced. Besides the difficulties with the temperature control described in the former Section, most notably the shielding and the grounding of the probe were poor. To account for this, the sample probe was rebuilt extensively. Fig. 5.7 shows the sample probe before [Fig. 5.7(a)] and after [Fig. 5.7(b)] improvements.

Before the rebuilding, the cap (yellow-green) only covered the measurement coil. The two glass capacitors and a large fraction of the connecting wires (black lines) were not shielded by the cap. Note that at the very beginning teflon capacitors were used, which were much larger than the glass capacitors. Quickly it was observed that these capacitors were blocked at low temperature, although they were designed to work at low temperatures. The replacing glass capacitors were smaller. This is the reason why the distance between the two plastic discs appears rather large and a big fraction of connecting wires was needed, which were not well shielded. The Cernox temperature sensor (purple) was also not shielded. Its connecting wires were only fixed inside an insulation sheath (thick olive stripe). A small copper plate (red disc) just below the first plastic disc connected the tuning capacitor to the ground. The core of the coaxial cable was not well shielded. All in all this layout led to a very poor shielding and grounding, resulting in a very poor signal-to-noise ratio.

During the improvements, the brass cap was extended until the first plastic disc. Now it covers not only the sample coil, but also the capacitors and the connecting wires and protects them. The core of the coaxial cable was pulled through the first plastic disc (together with the inner dielectric insulator of the coaxial cable), such that it is now also shielded by the new cap. Furthermore, the small copper plate (red) was enlarged. Only a circle around the matching capacitor was omitted in the new copper disc, to ensure insulation of this capacitor. All other elements, including all fixing brass tubes, are now connected to the ground via this new copper disc. The shield of the coaxial cable was directly soldered to the copper disc. The new elongated cap touches the new copper plate

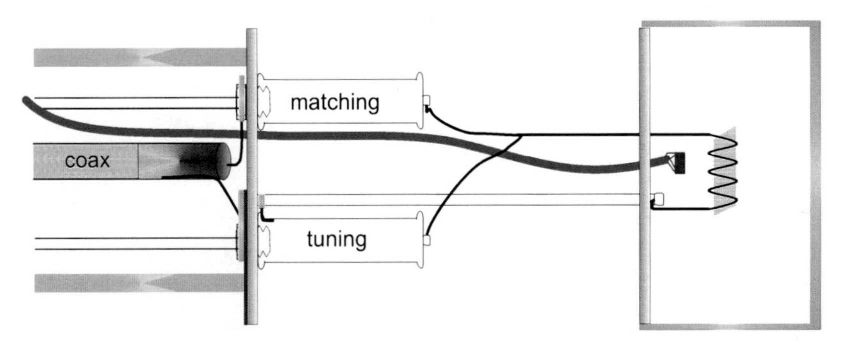


(a) **Sample probe before the improvement:** The cap (yellow-green) only covers the measurement coil. The Cernox temperature sensor (purple) and its connecting wires are not shielded.

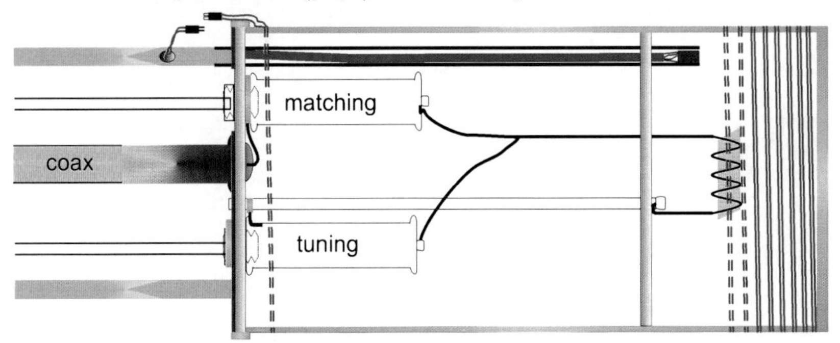


(b) **Sample probe after the improvement:** The cap is elongated, the temperature sensor and its wires are shielded from the rest of the probe. A copper plate (red) provides a good grounding of all elements and a cap heater (orange lines) was added.

**Figure 5.7:** Sketch of the 16 T sample probe before (upper panel) and after (lower panel) its improvement (for simplicity some fixing brass tubes which are not important for the improvements are omitted in these sketches): Fixing tubes of the sample probe are drawn as long, horizontally aligned, yellow-green squares. The grey, vertically aligned squares are plastic discs on which everything is fixed. The green thick square denotes the coaxial cable, the grey part of it is its shield, the black wire coming out of it denotes its core. Soldering lugs are sketched as small yellow-green squares. Soldered and connecting wires are thick black lines. The measurement coil is winded around the sample (grey rhomboid). The purple square is the Cernox temperature sensor.

which is a bit overlaying, such that the cap is well grounded, too. The insulation sheath containing the wires of the temperature sensor and the (newly installed) cap heater has been pulled through one of the hollow fixing tubes along the whole length of the sample probe. Another hollow tube (dark brown lines) has been put as a prolongation of this fixing tube, shielding the temperature sensor (purple square) and the insulation sheath (olive stripe) containing its connecting wires (thin blue stripes) from the environment. The wires of the heater exit the fixing tube before the beginning of the cap. A cap heater in form of a doubled resistance wire has been wound around the whole length of the cap (orange lines, only sketched at the beginning and the end) The cap heater is connected with the wires coming out of the fixing tube via two pairs of connectors.

The reported rebuilding of the lower part of the sample probe now guarantees good grounding and shielding as well as an optimized temperature control.

### 5.4.3 Sample Heating

High frequency pulses of a defined pulse length and power are used in the NMR pulse sequences. The applied power can be calculated as [38]:

$$P = \frac{2\omega V H_1^2}{\mu_0 Q}, \tag{5.1}$$

where $\omega = \gamma H_0$ is the NMR frequency, $V$ is the volume of the sample coil, $H_1$ is the magnetic field perpendicular to $H_0$, generated within the coil by the application of the pulses (see Fig. 2.4), $Q$ is the quality factor of the resonance circuit and $\mu_0 = 4\pi \times 10^{-7}\,\mathrm{Vs/Am}$ is the vacuum permeability. According to Eq. (2.44) the magnetic field generated in the coil by the application of a $\pi/2$ pulse of the length $t_p$ is:

$$H_1 = \frac{\pi/2}{\gamma t_p}. \tag{5.2}$$

For the $^{75}$As NMR measurements ($\gamma/2\pi = 7.2917\,\mathrm{MHz/T}$) on LiFeAs in $H_0 = 7\,\mathrm{T}$, typical pulse lengths of $2\mu$s were used. With these values, $H_1$ amounts to $0.017\,\mathrm{T}$. The coil dimensions correspond to the sample dimensions, which were $5.04\,\mathrm{mm} \times 2.24\,\mathrm{mm} \times 0.52\,\mathrm{mm}$. Assuming a typical quality factor of about 50, the applied power amounts to 17.3 W. For $^{7}$Li NMR measurements ($\gamma/2\pi = 16.5461\,\mathrm{MHz/T}$) in $H_0 = 4.5\,\mathrm{T}$, typical pulses were $1\,\mu$s long, which gives $H_1 = 0.015\,\mathrm{T}$ and an applied power of 19.6 W. Powers of such orders of magnitude are normally easy to compensate by the cooling power of usual $^{4}$He cryostats.

Conditions for the applied power change upon entering the superconducting state. Hence, attention has to be paid to select the right settings for pulse-length and power. The quality factor of a a parallel circuit of a coil of inductance $L$, a capacitor of capacity $C$ and a resistor with resistance $R$ is given by:

$$Q = R\sqrt{\frac{C}{L}}. \tag{5.3}$$

The inductance of a long coil with length $l$, cross section area $A$ and number of windings $N$ is:

$$L = \frac{\mu_0 \mu_r A N^2}{l}, \tag{5.4}$$

with $\mu_r$ being the relative permeability of the material inside the coil. Due to the perfect diamagnetism of superconductors, $\mu_r \to 0$ in the superconducting state and correspondingly $Q$ increases. Hence, according to Eq. (5.1) less power is needed to flip the nuclear spins in the superconducting state. Furthermore, due to their diamagnetism, superconductors shield radio frequencies. NMR signals from the superconducting state are therefore normally much weaker then the ones from the normal state. If one performs a pulse-length

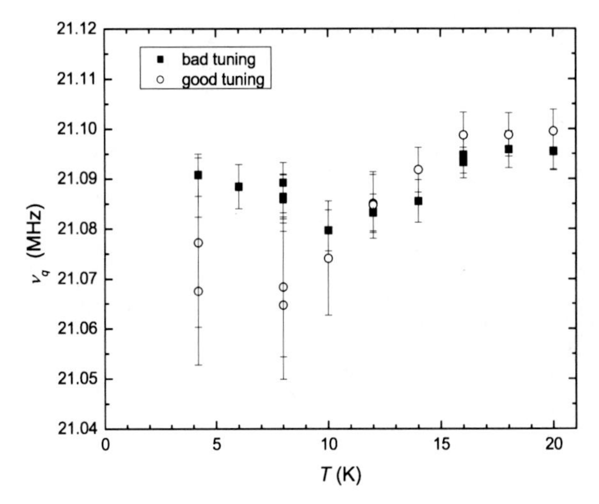


**Figure 5.8:** Effect of sample heating on the $^{75}$As NQR frequency of LiFeAs. The filled squares denote the measured NQR frequency with bad tuning conditions and sample heating for $T < 11\,\mathrm{K}$. The open dots show the corrected NQR frequency, measured with good tuning conditions. It seems however, that the 4.2 K point is still affected by sample heating.

measurement in the superconducting state just slightly below $T_c$, wrong pulses with too much power might be chosen because heat-excited normal-state signals will superimpose the much weaker signals from the superconducting state. The wrongly chosen high power either leads to unintentional flipping angles larger than 90° / 180° or to a heating of the sample.

Sample heating may also arise if the tuning conditions of the resonant circuit are poor and much power is reflected from the coil. To prevent any long air-contact for the very moisture- and air-sensitive LiFeAs sample, all measurements on the LiFeAs sample were done simultaneously in the same VTI with the same measurement coil. This implied that the same resonant circuit had to be used for a broad frequency range, ranging from 21.5 MHz for NQR-measurements[6] up to 62 MHz for $^{75}$As -NMR in 7.0494 T. With this set-up only very poor tuning conditions were obtained for low temperature NQR measurements, because the tuning changes drastically in the superconducting state. A lot of power was reflected by the electronic circuit. Because of the enhanced power reflectance of the circuit much more power had to be applied than what one would calculate from Eq. (5.1). The use of long pulses with high power resulted in a heating of the sample at low temperature and therewith to misleading results (see Fig. 5.8). It was therefore inevitable to take out the sample probe and install a supplementary capacitor of 100 pF

[6] The NQR measurements could be performed inside the magnet, since it was possible to drive the magnetic field down to zero in an oscillating mode. The good agreement between linewidths and frequencies of the measurements inside the 16 T cryostat with $H_0 = 0\,\mathrm{T}$ and other measurements in the NQR setup confirmed the absence of residual magnetic fields in the former case.

parallel to the tuning capacitor.[7] After that, very good tuning conditions were obtained also for NQR measurements at low temperature. A comparison between the measured NQR frequencies with bad and good tuning conditions at low temperature showed that the former sample heating significantly affected the NQR frequency (see Fig. 5.8). Since the NQR frequency was essential for the determination of the Knight shift (see Chapter 8), precise measurements of the NQR frequency were needed. The data obtained with the good tuning conditions seem to be quite reasonable for $T \geq 8\,\mathrm{K}$. The data at $4.2\,\mathrm{K}$ may however still be influenced by some sample heating effects.

[7] To minimize possible defects at the sample, it was kept under a helium gas flow atmosphere during the installation of the additional capacitor and everything was done as quickly as possible. A check of the NQR linewidth and frequency at room temperature after the installation of the additional capacitor proved that the sample was still undamaged.

# 6 NMR on $LaO_{1-x}F_xFeAs$ in the Normal State

In the following two chapters, NMR measurements on $LaO_{1-x}F_xFeAs$ with $x = 0.00$, 0.05, 0.075, and 0.10 will be presented. The presentation and discussion of the data will be separated into the normal state properties (this Chapter) and the nature of the superconducting state (following Chapter 7). A comparison to the data of other groups on the same compound and similar ones will be included.

As presented in the following, the normal state properties at high temperatures are dominated by a pseudogap-like decrease of static and dynamic NMR characteristics. At intermediate temperatures the dynamic properties depend on the doping level and thus on the proximity to the SDW instability. While antiferromagnetic spin fluctuations seem to boost the spin-lattice relaxation rate at low doping levels, no evidence for such fluctuations is found at optimal doping ($x = 0.1$). For this most thoroughly studied sample $LaO_{0.9}F_{0.1}FeAs$, a scaling of different NMR shifts and a scaling of different spin-lattice relaxation rates will be presented, suggesting a single spin fluid character and the lack of any $\vec{q}$-structure in the dynamic susceptibility, at least for this specific doping level.

## 6.1 Knight Shift - Static Susceptibility

Knight Shift measurements were only performed on samples with fluorine doping levels of $x = 0.05$ and $x = 0.10$.

### 6.1.1 Optimally-Doped $LaO_{0.9}F_{0.1}FeAs$

In the case of $LaO_{0.9}F_{0.1}FeAs$, an *ab*-oriented sample (see Chapter 5) was used for the Knight shift measurements. Only data for $H \parallel ab$ were taken. *The alignment of the sample and* $^{75}$*As-NMR measurements in a field of* $H_0 = 7.0494$ *T up to 300 K have already been performed by H.-J. Grafe and D. Paar and are published in [79].* In the course of this thesis, the $^{75}$As-NMR measurements were reproduced for some already existing temperature points and extended up to 480 K in $H = 7.0494$ T. Additionally, the $^{139}$La-NMR shift was measured in the temperature region from 100 K to 300 K in the same field of $H = 7.0494$ T. *More data points of the* $^{139}$*La-NMR shift were obtained by D. Paar, and G. M. Lang.*

The alignment process described in [79] worked well, yielding $^{75}$As-NMR resonance lines, that are well described by a Gaussian line shape (see Fig. 6.1). The resonance

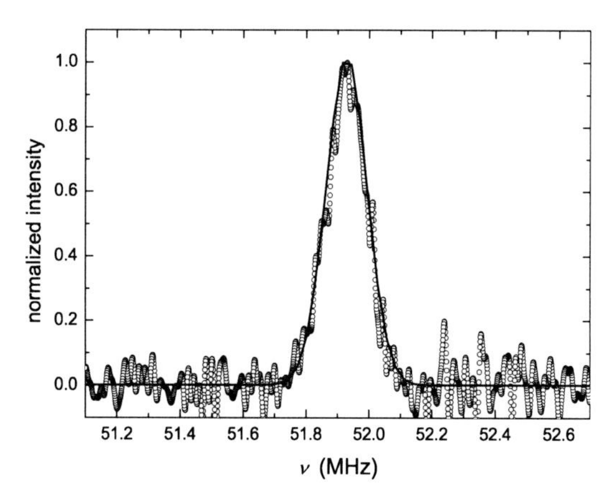


**Figure 6.1:** $^{75}$As-NMR resonance line of *ab*-aligned $LaO_{0.9}F_{0.1}FeAs$ in $H_0 = 7.0494\,\mathrm{T}$ for $H_0 \parallel ab$ and $T = 480\,\mathrm{K}$. The solid line is a Gaussian fit to the resonance line.

frequency $\nu_{ab,(\mathrm{Gauss})}$ could be extracted from a Gaussian fit in the whole temperature range. Additionally, the center of gravity of each resonance line was calculated:

$$\nu_{ab,(\mathrm{CoG})} = \frac{\sum_i \nu_i I_i(\nu_i)}{\sum_i I_i(\nu_i)}, \tag{6.1}$$

where $I_i(\nu_i)$ is the intensity at a certain frequency point $\nu_i$ of the resonance line. Both $^{75}\nu_{ab,(\mathrm{CoG})}$ and $^{75}\nu_{ab,(\mathrm{Gauss})}$ coincided within the small error bars. A similar procedure was applied to extract the $^{139}$La resonance frequencies $^{139}\nu_{ab}$.

With the obtained resonance frequencies $^{75}\nu_{ab}$ and $^{139}\nu_{ab}$, the $^{75}$As and $^{139}$La Knight shifts were extracted using Eq. (2.14), including also second order quadrupolar shift effects according to Eq. (2.39):

$$\nu_{ab} = \gamma H_0(1 + K_{ab}) + \frac{3\nu_q^2}{16\gamma H_0}. \tag{6.2}$$

For $^{75}$As, a temperature-independent quadrupole frequency of $\nu_q^{75} = 11\,\mathrm{MHz}$ and a vanishing quadrupole asymmetry parameter $\eta = 0$ were adopted. These values were deduced from a simulation of the powder spectrum of the sample before its alignment [79]. $\nu_q$ was simultaneously measured in $^{75}$As-NQR measurements on $LaO_{0.9}F_{0.1}FeAs$ [193] and agreed well with the value derived from the powder spectrum simulation. Eq. (6.2) assumes that the angle between the principal axis of the EFG with the largest eigenvalue, $V_{ZZ}$, and the direction of the magnetic field ($H \parallel ab$) is $\theta = 90°$, which implies that $V_{ZZ} \parallel c$. This has been confirmed during the alignment process [79]. In the case of $^{139}$La the already determined $\nu_q = 1.15\,\mathrm{Mhz}$ [193] was used to correct for the second order quadrupole shift.

Fig. 6.2 shows the temperature dependence of the $^{75}$As and $^{139}$La Knight shifts together with the $^{57}$Fe Knight shift and the macroscopic susceptibility. *The $^{57}$Fe Knight shift was determined by G. Lang and H.-J. Grafe on an $^{57}$Fe enriched powder sample [193]*

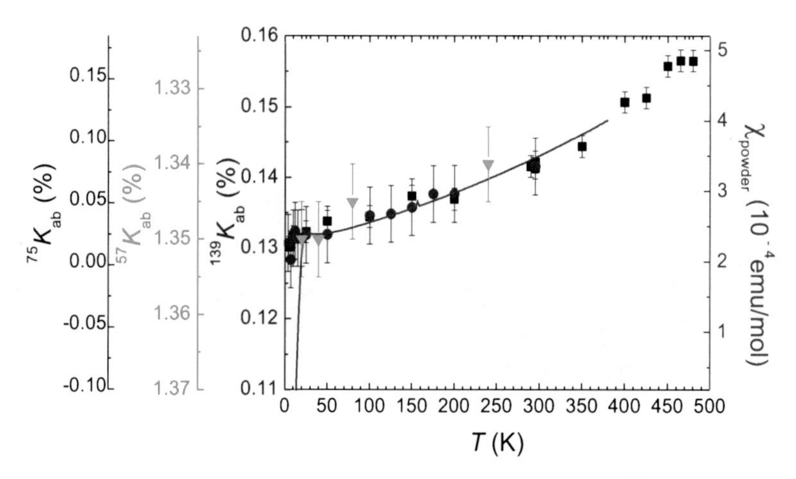


**Figure 6.2:** Comparison of the temperature-dependent Knight shifts of $^{75}$As (black squares), $^{139}$La (blue points) and $^{57}$Fe (green triangles) with the macroscopic susceptibility (solid red line) for $LaO_{0.9}F_{0.1}FeAs$ (published in [193]).

*and the macroscopic susceptibility was measured by N. Leps using a conventional SQUID magnetometer in a field of $H = 5$ T on a $LaO_{0.9}F_{0.1}FeAs$ powder sample [193].* Note that the scale for $^{57}K_{ab}$ is reversed, reflecting the negative hyperfine coupling constant for $^{57}$Fe.

At first glance it is already visible, that the macroscopic susceptibility scales with the microscopic Knight shifts down to $T_c$. This proves that the bulk susceptibility is not dominated by any Curie contribution due to paramagnetic impurities, and thus confirms the very high quality of the investigated samples.

The Knight shifts of all three nuclei decrease with decreasing temperature. This suppression of low energy spin excitations is a common feature observed in many pnictides [162, 166, 167, 226–230]. It has been compared to the pseudogap behavior found in cuprates (see for instance [31, 231, 232]). However, the expected broad maximum of the Knight shift at a crossover temperature $T_0$ in the high temperature regime, as reported e.g. for $YBa_2Cu_4O_8$ [231], has not been observed so far. The search for this broad maximum in the Knight shift was the reason to extend the measurements up to 480 K. No clear signature of such a "pseudogap peak" appears in the data plotted in Fig. 6.2. Only a slight flattening at around 500 K is visible. With the available setup, measurements at higher temperature were not possible. SQUID measurements suggest a degradation of the sample for temperatures higher than 480 K, which would possibly hinder reproducible measurements of the Knight shift at higher temperatures.

Following the analysis of [227, 229], the decrease of the Knight shift was fit with a pseudogap behavior of the form

$$K = A + B\exp(-\Delta_{PG}/k_BT)\,, \tag{6.3}$$

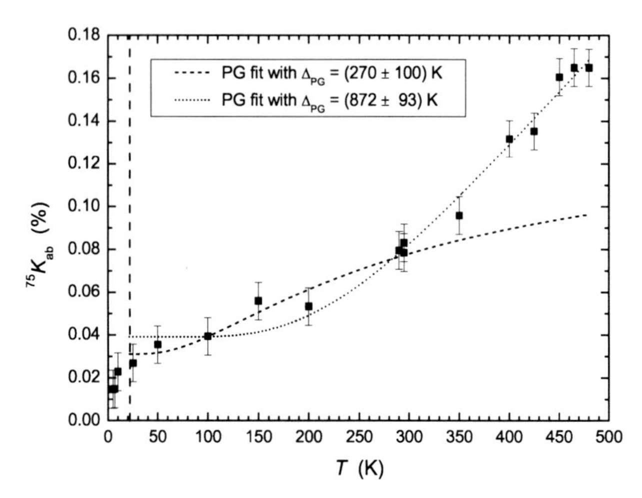


**Figure 6.3:** Temperature dependence of $^{75}K_{ab}$ of $LaO_{0.9}F_{0.1}FeAs$ measured at a field of $H_0 = 7.0494$ T. Lines are attempts of a pseudogap fit according to Eq. (6.3) for $T_c < T \leq 290$ K (dashed line) and $T_c < T \leq 480$ K (dotted line). The dashed vertical line denotes $T_c(H_0) \approx 22$ K.

with the constants $A$ and $B$ and the pseudogap $\Delta_{PG}$. Two fits for different temperature regions are shown in Fig. 6.3. The fit for $T_c < T \leq 290$ K (dashed line) gives a reasonable value for the pseudogap, but a very large error. Furthermore, it does not follow the data for $T > 300$ K. A fit over the whole paramagnetic temperature range (dotted line) can well describe the data, but results in $\Delta_{PG} = (872 \pm 93)$ K. This value is unrealistically high for an optimally-doped iron pnictide, considering the fact that pnictides are less correlated than cuprates. Possible reasons for the pseudogap-like behavior will be discussed in Section 6.1.3, after showing the data for $LaO_{0.95}F_{0.05}FeAs$ in Section 6.1.2. Additionally, also the spin-lattice relaxation rate measurements in Section 6.2 will be discussed in terms of a possible pseudogap behavior.

Another striking result of Fig. 6.2 is the fact, that all three Knight shifts do not only scale with the macroscopic susceptibility, but also with one another. This suggests that all three nuclei are probing the same component of the spin susceptibility, despite the multi-band character of the electronic structure (see Chapter 4). In cuprates, where a single-band model of the electronic structure is appropriate due to the localized character of the spin susceptibility (Cu $3d$), a single spin component has been proven by $^{63}$Cu and $^{17}$O NMR [31]. However, due to the multiband character of LaOFeAs one might expect that each band contributes differently to the spin susceptibility, with different selective hyperfine couplings to different bands, as it is the case in $Sr_2RuO_4$ for instance [233]. The scaling of all three Knight shifts with one another and with the macroscopic susceptibility suggests that if there are different hyperfine couplings to multiple orbitals in pnictides, their spin responses are nearly identical. This gives strong evidence for a single spin liquid

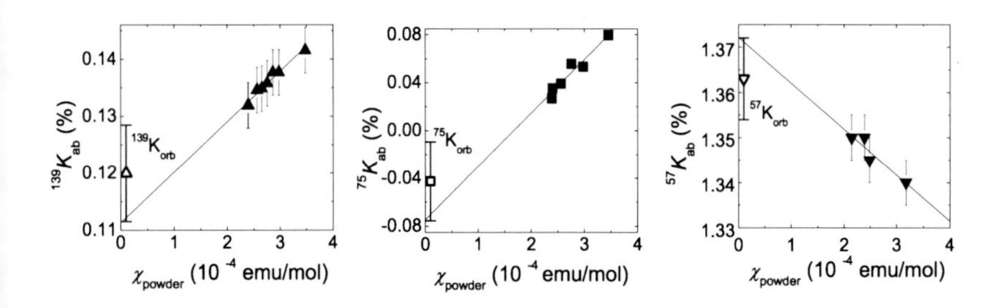


**Figure 6.4:** Knight shifts of $^{139}$La (left panel), $^{75}$As (center), and $^{57}$Fe (right panel) versus the macroscopic powder susceptibility with temperature as implicit parameter. The solid lines are linear fits. The range of possible orbital shifts is plotted next to each left vertical axis. These data are published in [79] and [193].

character of the electronic system, which may reflect the itinerant character of the Fe $3d$ electrons.

Note that the scaling of the macroscopic susceptibility with the measured Knight shifts is legitimate since both quantities contain a temperature-dependent contribution stemming from the macroscopic/local spin susceptibility and other temperature-independent parts stemming from van Vleck magnetism and diamagnetic responses. A shifting of the specific scales with respect to each other corresponds to a consideration of the temperature-independent parts $K_{orb}$ and $K_{dia}$, and $\chi_{VV}$ and $\chi_{dia}$, respectively as well as on the different hyperfine couplings $A_{hf}(\vec{q}=0)$.

If the macroscopic susceptibility is not covered by impurity contributions, as it is the case for $LaO_{0.9}F_{0.1}FeAs$, it is possible to extract the hyperfine coupling constants $A_{hf}(\vec{q}=0)$ by plotting the Knight shift versus the macroscopic susceptibility and fitting a linear dependence between these two, according to Eq. (2.21). The slope of the linear temperature dependence in this so-called Clogston-Jaccarino plot [234, 235] gives the hyperfine coupling constant at $\vec{q}=0$: $A_{hf}(\vec{q}=0)$. This is plotted in Fig. 6.4 for all three considered nuclei. Note that the powder susceptibility was used in this plot, neglecting the modest anisotropy between the $c$ and $ab$ direction.

The extracted hyperfine couplings amount to: $^{139}A_{hf} = 4.3(8)\,\mathrm{kOe}/\mu_B$, $^{75}A_{hf} = 25(3)\,\mathrm{kOe}/\mu_B$ and $^{57}A_{hf} = -5.7(14)\,\mathrm{kOe}/\mu_B$ [79, 193]. A rough estimation of the orbital shifts $K_{orb}$ along the $ab$ direction can also be extracted from this plot. It is given by the non-spin part of the susceptibility: $\chi_{ns} = \chi_{VV} + \chi_{dia}$, which can only be determined with large error bars. The lowest possible value of the non-spin part of the susceptibility is 0, while the upper bound is $2 \times 10^{-4}$ emu/mole. For higher values, the spin susceptibility would become negative below a certain temperature. This results in $^{139}K_{orb} = 0.12(1)\,\%$, $^{75}K_{orb} = -0.03(4)\,\%$ and $^{57}K_{orb} = 1.36(1)\,\%$ [79, 193]. Table 6.1 summarizes the extracted hyperfine couplings $A_{hf}^{ab}$ and orbital shifts $K_{orb}^{ab}$. The extracted values agree nicely with similar values reported by other groups. Terasaki *et al.* reported $^{57}A_{hf}(\vec{q}=0)/^{75}A_{hf}(\vec{q}=0) \simeq -0.38$ for $LaFeAsO_{0.7}$ [167]. This is of the same order of magnitude (and the same sign) as the value of $^{57}A_{hf}(\vec{q}=0)/^{75}A_{hf}(\vec{q}=0) \simeq -0.23$ which results out of our values. The same reference gives $^{57}K_{orb} = 1.425(1)\,\%$ [167], in

| | $^{139}$La | $^{75}$As | $^{57}$Fe |
|---|---|---|---|
| $A^{ab}_{hf}$ (kOe/$\mu_B$) | 4.3(8) | 25(3) | -5.7(14) |
| $K^{ab}_{orb}$ (%) | 0.12(1) | -0.03(4) | 1.36(1) |

**Table 6.1:** Hyperfine couplings and orbital shifts in $LaO_{0.9}F_{0.1}FeAs$ for $^{139}$La, $^{75}$As, and $^{57}$Fe [79, 193].

very good accord with our $^{57}K_{orb} = 1.36(1)\,\%$. For single crystals of undoped $BaFe_2As_2$ and undoped $CaFe_2As_2$, hyperfine coupling constants of $^{75}A_{hf} = 26.4(7)\,\mathrm{kOe}/\mu_B$ and $^{75}A_{hf} = 23\,\mathrm{kOe}/\mu_B$ were reported [224, 228]. The comparison with $^{75}A_{hf} = 25(3)\,\mathrm{kOe}/\mu_B$ for $LaO_{0.9}F_{0.1}FeAs$ indicates the relative robustness of the absolute value of the hyperfine coupling constant upon doping and structural changes and even among different pnictide families.

At first glance it is astonishing that the $^{57}$Fe hyperfine coupling constant does not possess the largest absolute value, as one would expect from the nature of the electronic structure which is predominantly governed by the presence of all Fe $3d$ bands at the Fermi level [119, 124, 189, 190, 236]. In contrast, the low value of $^{57}A_{hf}$ shows that $^{57}$Fe is a rather poor probe of the spin susceptibility at $\vec{q} = 0$. The negative sign indicates a dominant core polarization mechanism for $^{57}$Fe. One explanation of the low absolute value would be, that besides this core polarization mechanism, which is always large and negative; also a large, but positive contribution from the $4s$ electrons would be present, rendering the total hyperfine coupling $^{57}A_{hf} = {}^{57}A_{core} + {}^{57}A_{contact}$ (according to Equations (2.19) and (2.21)) rather weak. In principal, also hyperfine filtering effects could filter out the sensitivity of $^{57}$Fe at $\vec{q} = 0$. However, the spin-lattice relaxation rate measurements on $LaO_{0.9}F_{0.1}FeAs$ which will be discussed in Section 6.2.1 do not show any evidence for hyperfine filtering effects.

In contrast to $^{57}$Fe, $^{75}$As is a good probe for the spin susceptibility at $\vec{q} = 0$, since its hyperfine coupling constant $^{57}A_{hf} = 25\,\mathrm{kOe}/\mu_B$ is rather large. This may be due to the fact that the $^{75}$As ions are lying in the same FeAs layer as the iron ions that are responsible for the electronic properties. Each $^{75}$As ion is surrounded by four $^{57}$Fe ions and a strong hybridization between the As $p$ and the Fe $d$ bands is expected.

The low value of the hyperfine coupling constant of $^{139}$La compared to the one of $^{75}$As is not surprising, since $^{139}$La is located outside the FeAs layer and thus more weakly coupled to the dominant Fe $3d$ moments. One can use the extracted $^{139}A_{hf}$ to estimate the value of the iron moment in the SDW phase of undoped LaOFeAs, by using the value of the internal field in LaOFeAs at the $^{139}$La site $H_{int,x=0}(\mathrm{La}) = 2.5\,\mathrm{kOe}$ deduced from the splitting of the $^{139}$La-NMR spectrum in LaOFeAs [149] and by further assuming that the hyperfine coupling does not change significantly upon doping. This assumption is justified, since there exist no major structural changes between LaOFeAs and $LaO_{0.9}F_{0.1}FeAs$. It is furthermore confirmed by the absolute values of $^{75}A_{hf}$ for $LaO_{0.9}F_{0.1}FeAs$ ($25(3)\,\mathrm{kOe}/\mu_B$, see Table 6.1) and $BaFe_2As_2$ ($26.4(7)\,\mathrm{kOe}/\mu_B$, [228]), which are the same within error bars. If the $^{75}$As hyperfine coupling constant is that robust, no major changes are expected for the $^{139}$La hyperfine coupling constant, either.

The internal field at the $^{139}$La nucleus is caused by the iron moments $\mu$(Fe) via the hyperfine coupling $^{139}A_{hf}$:

$$H_{int}(\text{La}) = {}^{139}A_{hf} \times \mu(\text{Fe})\,. \tag{6.4}$$

With this relation one arrives at an iron moment of $0.58\,\mu_B$ per Fe atom [193]. The same procedure can be done for $^{75}$As. An internal field of $H_{int}(\text{As}) = 1.6\,\text{T} = 16\,\text{kOe}$ was found at the $^{75}$As site in undoped LaOFeAs by zero-field $^{75}$As-NMR at $T = 2.2\,\text{K}$ [237]. With

$$H_{int}(\text{As}) = {}^{75}A_{hf} \times \mu(\text{Fe}) \tag{6.5}$$

one obtains an ordered iron moment of $0.64\,\mu_B$ per Fe atom. Both values agree nicely with each other, further corroborating the single spin liquid character of the electronic system, which was discussed previously. Both nuclei, $^{139}$La and $^{75}$As are coupled to the same itinerant iron moments. The resulting average value of $\mu(\text{Fe}) = 0.61(3)\,\mu_B/\text{Fe}$ is somewhat bigger than what was determined by early measurements of neutron scattering and Mössbauer spectroscopy on LaOFeAs, which reported a magnetic moment of $0.25$-$0.35\,\mu_B/\text{Fe}$ [183–186, 238]. However, it perfectly agrees within error bars with more recent neutron scattering measurements which report a magnetic moment of $\mu(\text{Fe}) = 0.63(1)\,\mu_B/\text{Fe}$ [192]. Furthermore, it is consistent with DFT calculations, predicting that the size of the ordered moment is independent of the rare earth ion and with the observed ordered iron moment of $\mu(\text{Fe}) \approx 0.5\,\mu_B/\text{Fe}$ of PrOFeAs, which has been determined by two independent neutron powder diffraction measurements [194, 197]. For the other members of the "1111" family, the absolute value of the ordered iron moment is still under debate. Neutron powder diffraction and $\mu$SR measurements reported moments of $0.3\,\mu_B/\text{Fe}$ up to $0.8(1)\,\mu_B/\text{Fe}$ in CeOFeAs [136, 141]. Moments of $0.25(7)\,\mu_B/\text{Fe}$ and $0.9(1)\,\mu_B/\text{Fe}$ have been found by analyzing neutron data on NdOFeAs [195, 196]. The influence of the rare earth ion magnetism at low temperatures in these compounds and the lack of high quality single crystals for neutron studies further complicate the extraction of the correct ordered iron moment in these (RE)OFeAs systems [141]. On the other hand the moment deduced from Mösssbauer measurements [183, 184] depends on necessary assumptions regarding the average hyperfine field. These may falsify the determination of the absolute value of the moment, while there is a general agreement of Mössbauer measurements with the DFT prediction, that the moments should not strongly vary within the (RE)OFeAs series [141, 192].

All experimentally reported values so far are strongly reduced relative to theoretical expectations. This might reflect the largely itinerant nature of the SDW state. Several theoretical attempts were started to reconcile early DFT calculations predicting large magnetic moments with the experimentally observed small magnetic moments. Starting from a local moment picture (strong coupling, Mott physics) it is possible to describe the reduction of the magnetic iron moment by considering frustration effects between nearest neighbour and next-nearest neighbour exchange interactions [188, 239]. Another theoretical approach reduced the important electronic structure to one itinerant band and one more Mott-like localized band by including spin orbit interactions, strong hybridization between Fe $d$ and As $p$ bands and a compression of the lattice along $z$ [240]. The resulting spin component for relevant values of hybridizations is then reported to lie in between 0.2 and $0.6\,\mu_B/\text{Fe}$, which is very similar to the experimentally observed values. A third theoretical model suggests the existence of a large number of antiphase boundaries, stacking faults along $z$ direction and fluctuations of the SDW wavevector between $(\pi, 0)$ and $(0, \pi)$ and suggests that the combination of these effects will lead to a more *dynamic* SDW state with experimentally accessible small moments [191].

### 6.1.2 $LaO_{0.95}F_{0.05}FeAs$

The $^{75}$As Knight shift of $LaO_{0.95}F_{0.05}FeAs$ was measured from a powder sample in a magnetic field of $H = 7.0494\,T$, by determining the resonance frequency of the high frequency peak of the NMR powder spectrum, which corresponds to $H \parallel ab$ [79]. The correction of the resonance frequency for second order quadrupole effects according to Eq. (6.2) was complicated due to the line shape of the $^{75}$As-NQR spectrum of $LaO_{0.95}F_{0.05}FeAs$, showing two resonance lines with the corresponding quadrupole frequencies $\nu_{q,low} = 9.71$ Mhz and $\nu_{q,high} = 10.58$ MHz at room temperature [30]. The $^{75}$As Knight shifts was determined in three different ways, correcting for $\nu_{q,low}$, for $\nu_{q,high}$ as well as for their average $\nu_{q,mean}$. Additionally, the temperature dependence of $\nu_{q,low}$ and $\nu_{q,high}$ was taken into account, by fitting a linear decrease to the measured values of $\nu_{q,low}(T)$ and $\nu_{q,high}(T)$. This temperature dependence was also taken into account when calculating $\nu_{q,mean}(T)$.

Fig. 6.5 shows the temperature dependence of the thus obtained $^{75}$As Knight shifts for $LaO_{0.95}F_{0.05}FeAs$: $^{75}K_{ab,low}$, $^{75}K_{ab,high}$ and $^{75}K_{ab,mean}$. The error bars include uncertainties from the determination of the $^{75}$As-NMR resonance frequency and the $^{75}$As quadrupole frequency, as well as uncertainties from the linear fit to obtain $\nu_q(T)$. The absolute values

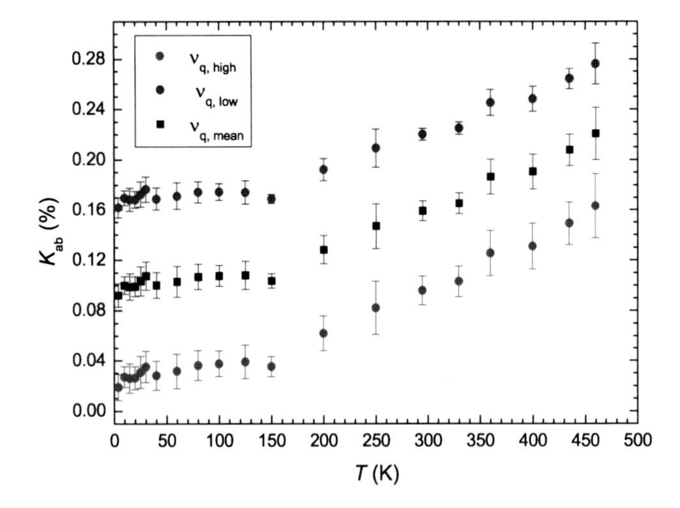


**Figure 6.5:** Temperature dependence of $^{75}K_{ab}$ for $LaO_{0.95}F_{0.05}FeAs$ determined in a field of $H = 7.0494$ T for temperatures up to 460 K. Three different values, determined by correcting the NMR resonance frequency for second order quadrupole corrections with $\nu_{q,high}(T)$ (red points), $\nu_{q,low}(T)$ (blue points) and $\nu_{q,mean}(T)$ (black squares), are shown.

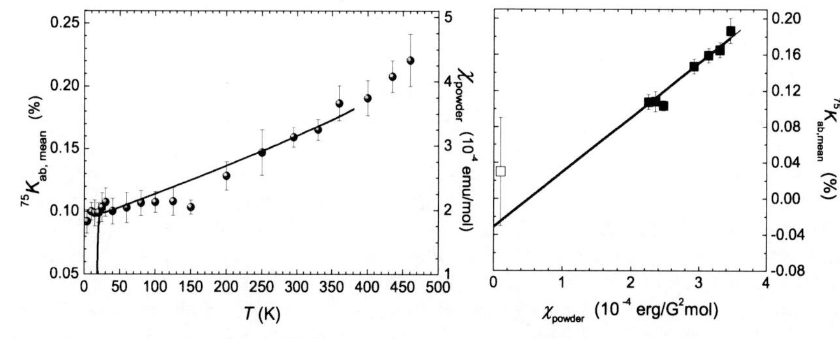


(a) Temperature dependence of $^{75}K_{ab,mean}$ (black dots) and the macroscopic susceptibility (solid line) (see also [181]).

(b) $^{75}K_{ab,mean}$ versus macroscopic susceptibility with temperature as implicit parameter. The solid line is a linear fit from $T = 100\,\mathrm{K}$ to $T = 360\,\mathrm{K}$. The averaged orbital shift is plotted next to the left vertical axis.

**Figure 6.6:** Scaling of the averaged knight shift $^{75}K_{ab,mean}$ and the macroscopic powder susceptibility $\chi_{powder}(T)$ for $LaO_{0.95}F_{0.05}FeAs$.

of the Knight shifts determined in such a way differ strongly from one another, ranging from $^{75}K_{ab,low}(295\,K) = 0.22\,\%$ to $^{75}K_{ab,high}(295\,K) = 0.09\,\%$ and being incompatible with each other inside their corresponding error bars. However, their overall temperature dependence is the same. They all decrease with decreasing temperature with the same slope.

The overall temperature dependence of the $^{75}$As Knight shift can be nicely scaled with the macroscopic susceptibility measured in a field of $H = 1\,\mathrm{T}$ [see Fig. 6.6(a)]. As in the case of $LaO_{0.9}F_{0.1}FeAs$ this reflects the high quality of the investigated samples since it proves that the macroscopic powder susceptibility is not governed by impurity contributions. A plot of $^{75}K_{ab,mean}$ versus the macroscopic susceptibility gives a linear dependence. This is plotted in Fig. 6.6(b). Different linear fits for different temperature regions, ranging from 60 K - 360 K to 200 K - 360 K were performed, They all yielded reasonable results. Fig. 6.6(b) shows a fit in the temperature interval from 100 K to 360 K. The averaged hyperfine coupling constant of all the fits amounts to $^{75}A_{hf} = 33(2)\,\mathrm{kOe}/\mu_B$ and the mean orbital shift to $^{75}K_{orb} = 0.03(6)\,\%$. These values nicely agree with the ones extracted for $LaO_{0.9}F_{0.1}FeAs$ (see Table 6.1) if one considers the difficulty in extracting the Knight shift of $LaO_{0.95}F_{0.05}FeAs$ due to the complicated $^{75}$As-NQR spectrum.

### 6.1.3 Discussion

Several theoretical approaches have been made to discuss the decrease of the spin susceptibility with decreasing temperature in the context of antiferromagnetic fluctuations. They include the consideration of antiferromagnetic fluctuations of local magnetic moments within a localized description [241] as well as within an itinerant approach considering the nesting between hole and electron pockets, which boosts magnetic correlations at $\vec{Q} = (\pi, \pi)$ [242, 243].

Another theoretical approach discusses the preformation of Cooper pairs driven by attractive excitonic interactions which may arise due to the large polarizability of the anions and the peculiar lattice structure as a reason for the decrease of the susceptibility with decreasing temperature [244, 245].

Effects of the electronic band structure such as an unusual temperature dependence of the density of states might also lead to a decreasing spin susceptibility. The interpretation of the linear temperature dependence of the susceptibility in terms of a pseudogap behavior similar to the one established for the case of cuprates has been discussed mostly in early NMR papers [79, 149, 226]. However, the doping-independent slope of the macroscopic susceptibility over the whole doping range [181] would require the pseudogap to exist over the whole doping range, contrarily to what is observed for cuprates [246]. Furhermore, as will be shown in Section 6.2.1, it is not possible to fit the decrease of $(T_1T)^{-1}$ consistently with a pseudogap function over the whole paramagnetic temperature range.

### 6.1.4 Summary

$^{75}$As NMR Knight shift data have been presented for underdoped $LaO_{0.95}F_{0.05}FeAs$ and optimally-doped $LaO_{0.9}F_{0.1}FeAs$. A robust feature of both data sets is the decrease of the Knight shift with decreasing temperature in the whole paramagnetic temperature range. Although a pseudogap behavior was suggested by several authors, it seems not to be very likely since the expected broad pseudogap peak was not observed in the measured temperature range which ranges up to 480 K. Different theoretical approaches were shortly mentioned in Section 6.1.3.

For the optimally-doped sample, the $^{75}$As NMR Knight shift was compared to $^{57}$Fe and $^{139}$La Knight shifts. A scaling of all three shifts together with the macroscopic susceptibility suggests that despite the existence of multiple bands at the Fermi energy, all nuclear spins are probing the same spin degree of freedom and the magnetic properties can therefore be well described within a single spin fluid model. Hyperfine coupling constants could be extracted and understood for all considered nuclei. The magnitude of the ordered magnetic moment in the SDW ordered state of LaOFeAs could be extracted. It amounts to $0.61(3)\,\mu_B$ per Fe atom which agrees nicely with recent neutron scattering experiments [192].

## 6.2 Spin-Lattice Relaxation Rate - Dynamics

Measurements of $^{75}$As-NMR spin-lattice relaxation rate were performed on samples with different fluorine doping levels and in slightly different temperature regimes: $x = 0$ ($T = 136 - 300\,\text{K}$), $x = 0.05$ ($T = 1.7 - 460\,\text{K}$), $x = 0.075$ ($T = 10 - 300\,\text{K}$) and $x = 0.1$ ($T = 2 - 10\,\text{K}$ and $T = 295 - 480\,\text{K}$). *Additional measurements of $^{75}$As-NMR spin-lattice relaxation in a powder sample with $x$=0.1 were done in the temperature range from 4.2 to 295 K by D. Paar.* In this optimally-doped sample $LaO_{0.9}F_{0.1}FeAs$, also the $^{139}$La-NMR spin-lattice relaxation rate was measured in between 100 and 300 K. *Some more* *data points at lower temperatures were taken by D. Paar. The $^{57}$Fe-NMR spin-lattice relaxation in $LaO_{0.9}F_{0.1}FeAs$ was obtained by G. Lang and H.-J. Grafe.*

The discussion of the data will again start with the optimally-doped sample $LaO_{0.9}F_{0.1}FeAs$. Relaxation rates of different nuclei in this sample will be compared with each other. Subsequently the doping evolution of the temperature dependence of the $^{75}$As-NMR spin-lattice relaxation rate will be analysed, especially in terms of the role of spin fluctuations in superconducting samples. This discussion will be split into a qualitative (Section 6.2.3.1) and a quantitative (Section 6.2.3.2) part.

For a discussion of the data it is important to remind that according to Eq. (2.53), the spin-lattice relaxation rate in correlated materials is proportional to the $\vec{q}$-dependent dynamic susceptibility:

$$\frac{1}{T_1T} \propto \gamma^2 \sum_{\vec{q}} |A_\perp(\vec{q})|^2 \frac{\chi''_\perp(\vec{q},\omega_L)}{\omega_L}. \tag{6.6}$$

### 6.2.1 Optimally-Doped $LaO_{0.9}F_{0.1}FeAs$

Measurements of $^{75}$As-NMR spin-lattice relaxation rate were done in an *ab*-oriented sample of $LaO_{0.9}F_{0.1}FeAs$ for $H_0 \parallel ab$ at low temperature ($T < 10\,$K, see Section 7.1) and high temperature ($T > 295\,$K) in a field of $H_0 = 7.0494\,$T. They were added to already existing $T_1$ data in the "normal temperature" region (4.2 K - 295 K), measured in the same magnetic field of 7.0494 T on the $H \parallel ab$ peak of the powder spectrum of a polycrystalline sample of $LaO_{0.9}F_{0.1}FeAs$ by D. Paar [79].

Fig. 6.7 shows the temperature dependence of the $^{75}$As-NMR $(T_1T)^{-1}$ of both investigated samples in $H_0 = 7.0494\,$T. A generally good agreement is observed, reflecting the reproducibility of the data and thus the high quality of the samples. From the highest measured temperature, 480 K, down to the onset of superconductivity at $T_c(H_0) \approx 22\,$K, $(T_1T)^{-1}$ decreases with decreasing temperature. This behavior has been widely observed in optimally and overdoped pnictides [148, 149, 167, 168, 248]. It stands in contrast to a simple Fermi liquid behavior (in this case, a constant $(T_1T)^{-1}$ would be expected in the normal state), but is reminiscent of a pseudogap-like behavior similar as the Knight shift data.

However, no pseudogap peak appears up to 480 K. In the cuprates, a broad maximum, named pseudogap peak, around the suspected pseudogap temperature $T^*$ has been observed in $^{63}$Cu-NMR [23, 31] and has been associated with the opening of a pseudo spin gap. The absence of this peak up to relative high temperatures in the measured $^{75}$As $(T_1T)^{-1}$ questions the pseudogap scenario, although hyperfine filtering effects could hide the pseudogap peak, as in the case of $^{17}$O $(T_1T)^{-1}$ in cuprates [23, 31, 249]. However, as shown in Fig. 6.10(a) and discussed later in this Section, the $^{75}$As-NMR $(T_1T)^{-1}$ of LaOFeAs, $LaO_{0.95}F_{0.05}FeAs$ and $LaO_{0.925}F_{0.075}FeAs$ show a pronounced increase with decreasing temperature, indicative of antiferromagnetic fluctuations. This proves, that $^{75}A_{hf}(\vec{q})$ is sensitive to antiferromagnetic fluctuations. Furthermore, although the hyperfine coupling constant of iron $^{57}A_{hf}(\vec{q})$ is supposed to be largest at $\vec{q} = (\pi,\pi)$ [167], the $^{57}$Fe $(T_1T)^{-1}$ in $LaO_{0.9}F_{0.1}FeAs$ decreases in the same way as the one of $^{75}$As, without any hint for a pseudogap peak (see Fig. 6.9). To completely rule out the pseudogap scenario proposed by several references, we followed their approach and fitted $(T_1T)^{-1}$ with a pseudogap function of the form [149, 166, 168, 230]:

$$\frac{1}{T_1T} = A + B\exp(-\Delta_{PG}/T)\,. \tag{6.7}$$

Two of the best fits are plotted in Fig. 6.7. A fit from $T_c < T \leq 290\,\text{K}$ results in a pseudogap value of $\Delta_{PG} = (163 \pm 23)\,\text{K}$ (dashed line). This value is similar to $\Delta_{PG} = (172 \pm 17)\,\text{K}$ obtained for $LaO_{0.89}F_{0.11}FeAs$ in the same temperature region [149]. However, this fit can not describe the new data at higher temperatures. Another fit for the whole measured paramagnetic temperature range $T_c < T \leq 480\,\text{K}$ yields $\Delta_{PG} = (473 \pm 46)\,\text{K}$ (dotted line). Yet this fit does not follow the data at low temperatures. In conclusion, no satisfying pseudogap fit could be found for the entire temperature range. Together with the unrealistically high value of $\Delta_{PG} = (872 \pm 93)\,\text{K}$ deduced from a pseudogap fit on the Knight shift data of $LaO_{0.9}F_{0.1}FeAs$ (see Fig. 6.3), this makes any pseudogap interpretation of the decreasing low energy spin excitations very unlikely and suggests that the physics behind the decrease of $K$ and $(T_1T)^{-1}$ are of a different nature than in the cuprates.

Interestingly, the $^{75}$As $(T_1T)^{-1}$ in the whole accessed paramagnetic regime can be perfectly described by a linear temperature dependence of the form:

$$\frac{1}{T_1T} = a + bT\,. \tag{6.8}$$

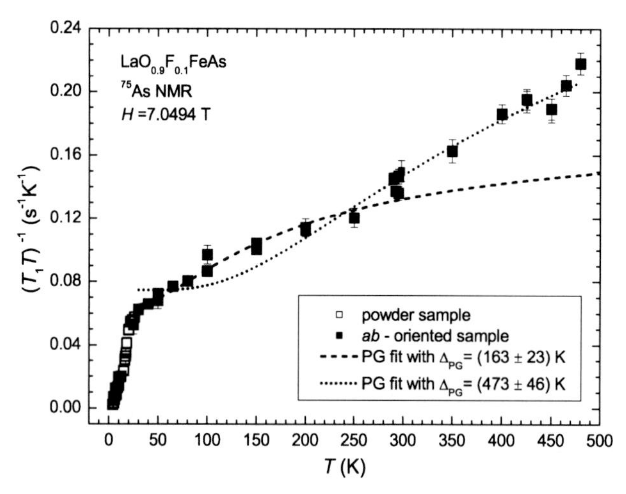


**Figure 6.7:** Temperature dependence of the $^{75}$As-NMR $(T_1T)^{-1}$ for $H \parallel ab$ measured at a field of $H = 7.0494\,\text{T}$ on a powder sample (open squares, published in [79]) and an $ab$ oriented sample (filled squares) of $LaO_{0.9}F_{0.1}FeAs$. Solid lines are attempts of a pseudogap fit according to Eq. (6.7) for $T_c < T \leq 290\,\text{K}$ (dashed line) and $T_c < T \leq 480\,\text{K}$ (dotted line) (published similarly in [247]).

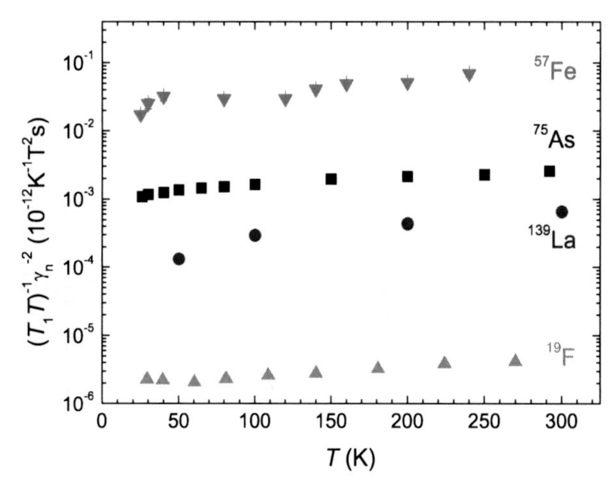


**Figure 6.8:** Temperature dependence of $^{57}$Fe, $^{75}$As, $^{139}$La, and $^{19}$F $(T_1T)^{-1}$ divided by the square of the corresponding nuclear gyromagnetic ratio $\gamma$. Data for $^{19}$F are reproduced from [226]. For $^{57}$Fe, $^{75}$As, and $^{139}$La, $T_1^{-1}$ along $ab$ is plotted, while for $^{19}$F it is $1/T_1^{iso}$.

The corresponding fit with $a = (0.052 \pm 0.002)\,\mathrm{s^{-1}K^{-1}}$ and $b = (3.2 \pm 0.1) \times 10^{-4}\,\mathrm{s^{-1}K^{-2}}$ is plotted in Fig. 6.15. The microscopic origin of this linear decrease is unclear. Similarly to the Knight shift, it might be related to temperature-dependent changes in the electronic density of states.

Fig. 6.8 shows the temperature dependence of $(T_1T)^{-1}$ of all measured nuclei, as well as $^{19}$F-NMR data from another group [226]. The data were scaled by the square of the corresponding nuclear gyromagnetic ratio $\gamma$, such that a quantitative comparison can be made between the spin-lattice relaxation rates of the four different nuclei. The absolute values of $(T_1T\gamma^2)^{-1}$, spreading over three orders of magnitude, reflect the distance of the corresponding nuclei to the iron plane. $(T_1T\gamma^2)^{-1}$ is largest for the $^{57}$Fe nuclei, while for $^{139}$La, and $^{19}$F, which lie outside the FeAs planes, the relaxation is much slower. This observation is not inconsistent with the Knight shift data which showed that due to the relative low hyperfine coupling constant $^{57}A_{hf}(\vec{q} = 0)$, $^{57}$Fe is a poor probe of the uniform susceptibility at $\vec{q} = 0$. According to Eq. (6.6) $T_1^{-1}$ probes the dynamic susceptibility in the whole $\vec{q}$-space and not only at $\vec{q} = 0$ and so one finds the expected situation: the $^{57}$Fe nuclei experience the fastest relaxation, which means that they are strongest coupled to the electronic spins and then the strength of the relaxation correlates with the distance to the iron plane. Correspondingly, one finds the interesting ratios: $^{75}(T_1T\gamma^2)/^{57}(T_1T\gamma^2) \approx 20 - 30$, while $(^{57}A_{hf})^2(\vec{q} = 0)/(^{75}A_{hf})^2(\vec{q} = 0) \approx 0.05$. This seeming discrepancy can be explained within two scenarios. Either there exists a strong $\vec{q}$-dependence of the dynamic susceptibility $\chi''(\vec{q}, \omega_L)$ and due to hyperfine filtering effects the iron probes fluctuations at $\vec{q} \neq 0$ better than the arsenic does and vice versa at $\vec{q} = 0$;

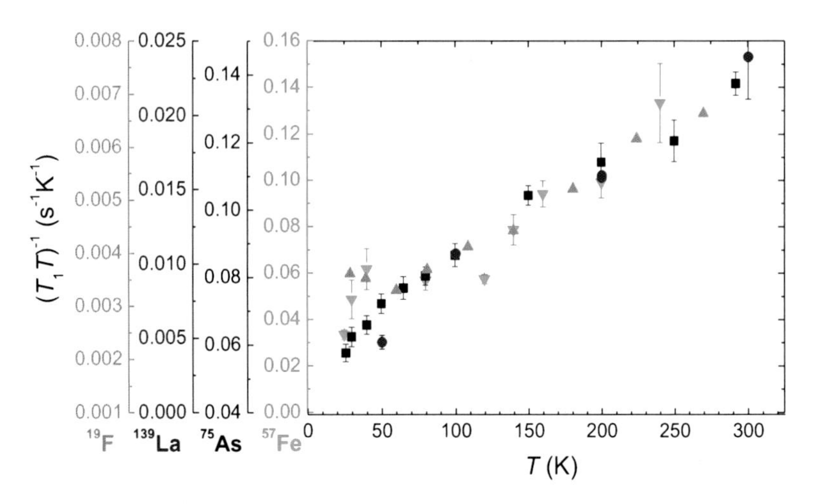


**Figure 6.9:** Same data as in Fig. 6.8, now scaled to each other and without the normalization by $\gamma^2$. Again, data for $^{19}F$ have been taken from [226] and $T_1^{-1}$ along *ab* is plotted $^{57}Fe$, $^{75}As$, and $^{139}La$, while $1/T_1^{iso}$ is plotted for $^{19}F$ [193].

or the most relevant scattering process is simple quasi particle scattering and the different ratios stem from different hyperfine coupling channels of the $^{57}Fe$ nuclear moments to the single spin fluid.

The first possibility of a strong $\vec{q}$-dependence is very unlikely, since the overall temperature dependence of the (in absolute values very different) spin-lattice relaxation rates is the same for all four nuclei. This scaling is depicted in Fig. 6.9. It shows that spin excitations are suppressed simultaneously across the whole $\vec{q}$-space as temperature is decreased. Any $\vec{q}$-dependence of the dynamic susceptibility $\chi''(\vec{q}, \omega_L)$ should show up in the comparison of spin-lattice relaxation rates of different nuclei, since the complicated multiband electronic structure should lead to different $\vec{q}$-dependences of each hyperfine coupling, which will then filter out different $\vec{q}$-regions of the dynamic susceptibility $\chi''(\vec{q}, \omega_L)$. This seems not to be the case in $LaO_{0.9}F_{0.1}FeAs$, as all four different spin-lattice relaxation rates scale nicely with each other in the whole paramagnetic regime.

The easiest interpretation is then that simple quasi particle scattering is the main cause for spin-lattice relaxation, which renders the existence of antiferromagnetic correlations in $LaO_{0.9}F_{0.1}FeAs$ unlikely. The rather weak hyperfine coupling at $\vec{q} = 0$ which was found for $^{57}Fe$ in static NMR measurements was interpreted as the sum of a large, negative core polarization term and a second large, but positive Fermi contact term: $^{57}A_{hf}(\vec{q} = 0) =^{57} A_{core}(\vec{q} = 0) +^{57} A_{contact}(\vec{q} = 0)$. In the case of simple quasi particle scattering, these terms enter Eq. (6.6) as the sum of their squares, $(^{57}A_{core})^2 + (^{57}A_{contact})^2$, rather than the square of their direct sum [250], leading to the observed fast spin-lattice relaxation of the $^{57}Fe$ nuclei without contrasting the weak hyperfine coupling at $\vec{q} = 0$. Note however that some fluctuations in certain $\vec{q}$-regions cannot be ruled out within this simple analysis.

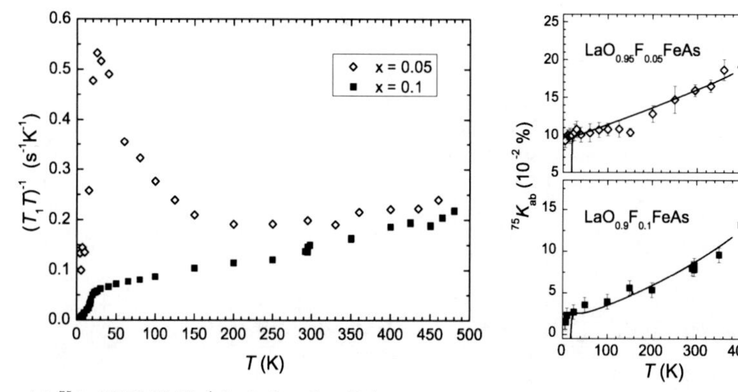


(a) $^{75}$As-NMR $(T_1T)^{-1}$ for $LaO_{0.95}F_{0.05}FeAs$ (open rhombi) and $LaO_{0.9}F_{0.1}FeAs$ (filled squares).

(b) $^{75}K_{ab}$ of $LaO_{0.95}F_{0.05}FeAs$ (upper panel) and $LaO_{0.9}F_{0.1}FeAs$ (lower panel) together with their corresponding macroscopic susceptibilities (solid lines).

**Figure 6.10:** Comparison of dynamic (left panel) and static (right panel) NMR properties of $LaO_{0.95}F_{0.05}FeAs$ and $LaO_{0.9}F_{0.1}FeAs$ in a field of $H_0 = 7.0494\,$T. The right panel is published similarly in [181]. For $^{75}K_{ab}$ of $LaO_{0.95}F_{0.05}FeAs$ the mean value has been plotted (see discussion in Section 6.1.2).

### 6.2.2 Underdoped $LaO_{0.95}F_{0.05}FeAs$

Before analysing the overall doping dependence of the spin-lattice relaxation rate from $x = 0.0$ to $x = 0.1$, this Section will start with a short discussion of the $^{75}$As $(T_1T)^{-1}$ of $LaO_{0.95}F_{0.05}FeAs$ in comparison to $(T_1T)^{-1}$ of $LaO_{0.9}F_{0.1}FeAs$ and will compare these dynamic NMR properties with the corresponding static ones. Fig. 6.10(a) compares the temperature dependence of the $^{75}$As-NMR $(T_1T)^{-1}$ of $LaO_{0.95}F_{0.05}FeAs$ (open rhombi) with the one of $LaO_{0.9}F_{0.1}FeAs$ (filled squares). From high temperature down to room temperature, $(T_1T)^{-1}$ of $LaO_{0.95}F_{0.05}FeAs$ decreases similarly to $(T_1T)^{-1}$ of $LaO_{0.9}F_{0.1}FeAs$. Below room temperature, however, the spin-lattice relaxation rate of $LaO_{0.95}F_{0.05}FeAs$ exhibits a Curie-Weiss-like increase with decreasing temperature down to 25 K, in sharp contrast to the temperature dependence of $(T_1T)^{-1}$ in $LaO_{0.9}F_{0.1}FeAs$. This is a signature of a slowing down of antiferromagnetic fluctuations, which seem to precede the superconducting ground state. Recall that $LaO_{0.95}F_{0.05}FeAs$ does not show a magnetic phase transition, but becomes superconducting at $T_c = 20\,$K ($T_c(H_0) = 17\,$K). Indeed, it is the first superconducting sample of the $LaO_{1-x}F_xFeAs$ series and thus right on the border between a magnetic and a superconducting ground state (see Fig. 4.6). Such an enhancement of $(T_1T)^{-1}$ is only found in underdoped superconducting samples of $LaO_{1-x}F_xFeAs$ and at low temperature (see also discussion of Fig. 6.11(a) and Section 6.2.3.2). It indicates the occurrence of antiferromagnetic spin fluctuations in the underdoped regime and demonstrates the important interplay between magnetism and superconductivity in the La1111 family.

Fig. 6.10(b) shows the behavior of the Knight shift, which, in contrast to the very different behavior of $(T_1T)^{-1}$, decreases with decreasing temperature in the underdoped $LaO_{0.95}F_{0.05}FeAs$ as well as in the optimally-doped $LaO_{0.9}F_{0.1}FeAs$. The Knight shift, which is a measure of the static susceptibility at $\vec{q} = 0$, decreases in both samples. The spin-lattice relaxation rate $(T_1T)^{-1}$ is also sensitive for $\vec{q} > 0$ contributions and is therefore more sensible for antiferromagnetic fluctuations, which give rise to an enhanced dynamic susceptibility at $\vec{q} = \pi$. The enhancement of $(T_1T)^{-1}$ in $LaO_{0.95}F_{0.05}FeAs$ shows that these low energy antiferromagnetic spin fluctuations are enhanced in the underdoped region, while they are suppressed in the optimally-doped $LaO_{0.9}F_{0.1}FeAs$.

## 6.2.3 Overall Doping Dependence - Role of Spin Fluctuations

### 6.2.3.1 Qualitative Discussion

The question arises whether these spin fluctuations favour or hinder superconductivity. To answer this question, Fig. 6.11(a) shows the temperature dependence of the $^{75}$As-NMR $(T_1T)^{-1}$ of all measured samples, with a fluorine doping level ranging from $x = 0.0$ to $x = 0.1$. The undoped sample (dark red triangles) is the only investigated sample which undergoes a magnetic phase transition at $T_N = 138\,\text{K}$. Its spin-lattice relaxation rate divided by temperature, $(T_1T)^{-1}$, increases from room temperature on with decreasing temperature and diverges towards its magnetic ordering temperature [see inset of

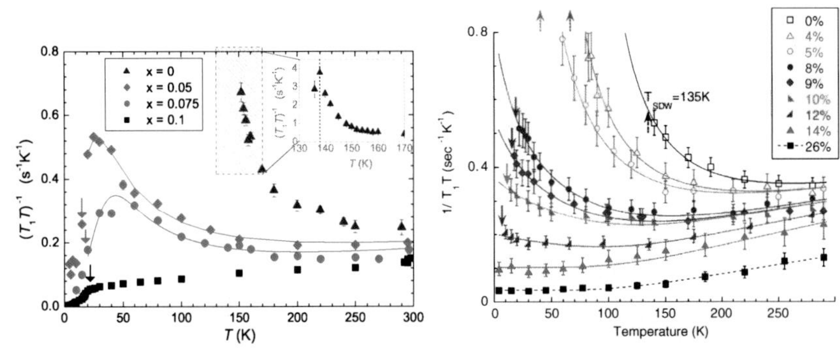


(a) $^{75}$As $(T_1T)^{-1}$ of $LaO_{1-x}F_xFeAs$ for various fluorine concentrations $x$, measured for $ab \parallel H_0 = 7.0494\,\text{T}$. The inset shows the complete divergence of $(T_1T)^{-1}$ of LaOFeAs near its magnetic ordering temperature $T_N = 138\,\text{K}$ (vertical dark red line).

(b) $^{75}$As $(T_1T)^{-1}$ of $Ba(Fe_{1-x}Co_x)_2As_2$ for various cobalt concentrations $x$, measured in $H_0 \approx 7.74\,\text{T}$. Upward arrows denote $T_N$, downward arrows denote $T_c$ (reproduced from [230]).

**Figure 6.11:** Comparison of the spin dynamics of $LaO_{1-x}F_xFeAs$ (a) and $Ba(Fe_{1-x}Co_x)_2As_2$ (b) in form of their temperature-dependent $^{75}$As $(T_1T)^{-1}$ for various doping levels $x$, including magnetic samples [$x = 0$ (a) and $x = 0, 4, 5\%$ (b)], superconducting samples [$x = 0.05, 0.075, 0.1$ (a) and $x = 8, 9, 10, 12, 14\%$ (b)] and a non-superconducting sample (x = 26%) in the case of $Ba(Fe_{1-x}Co_x)_2As_2$. Same scales are chosen for a better comparison.

Fig. 6.11(a)]. The temperature dependence of this increase and a similar absolute value of $(T_1T)^{-1}$ in the ordered state slightly below $T_N$ suggest a second-order-like phase transition to the magnetically-ordered state, in contrast to $BaFe_2As_2$, where only a slight increase of $(T_1T)^{-1}$ above $T_N$ and a discontinuous decrease of $(T_1T)^{-1}$ across $T_N$ is observed [251]. The behavior of the 5% F-doped sample (blue rhombi) was discussed in the former paragraph. The sample with a fluorine concentration of $x = 0.075$ (green circles) shows a very similar temperature dependence: after a slight decrease between 300 K and 220 K, it increases with decreasing temperature. This Curie-Weiss-like behavior in both underdoped samples, $x = 0.05$ and $x = 0.075$, indicates the existence of antiferromagnetic fluctuations which precede the superconducting state. The decrease of the absolute values of $(T_1T)^{-1}$ with increasing doping reflects a reduction of the antiferromagnetic correlation strength with doping. The observation of the persistence of antiferromagnetic fluctuations in underdoped superconducting samples with $0.05 \leq x \leq 0.075$ is consistent with the observation of an inflection point at around 150 K and a low-temperature upturn in resistivity measurements on these compounds. These anomalous features were related to a carrier localization due to remnant spin fluctuations [180].

The temperature dependence of $(T_1T)^{-1}$ in the optimally-doped sample ($x = 0.1$, black squares) does not show any hint for antiferromagnetic fluctuations any more. As already discussed in Section 6.2.1 it decreases linearly with decreasing temperature in the whole paramagnetic regime. The entirety of these data suggest that antiferromagnetic correlations remnant of the antiferromagnetically ordered SDW state remain present in the underdoped superconducting regime, but the highest superconducting transition temperature $T_c$ is only reached as soon as these fluctuations are completely suppressed (at least on the time scale of NMR measurements). *Thus, spin fluctuations seem to compete with superconductivity.*

This observation stands in contrast to measurements of $^{75}$As-NMR $(T_1T)^{-1}$ of various samples of $Ba(Fe_{1-x}Co_x)_2As_2$ [230], plotted in Fig. 6.11(b). The optimally-doped sample of this family ($x = 0.08$, dark blue circles), shows a strong enhancement of antiferromagnetic spin fluctuations, indicated by the strong upturn of $(T_1T)^{-1}$. This increase is also observable in slightly overdoped samples up to $x \leq 0.12$. Antiferromagnetic spin fluctuations are only suppressed completely in heavily overdoped samples with low or vanishing $T_c$ [see also phase diagram of $Ba(Fe_{1-x}Co_x)_2As_2$ in Fig. 6.12(b)]. The authors of [230] conclude that a modest enhancement of antiferromagnetic spin fluctuations, just strong enough to not cause a SDW ordered state, is needed to obtain the maximal $T_c$ in $Ba(Fe_{1-x}Co_x)_2As_2$ and that a further suppression of spin fluctuations goes along with a reduction of $T_c$. It is known from cuprate superconductors, that antiferromagnetic spin fluctuations can provide attractive interactions similar to lattice vibrations in BCS superconductors and thus mediate the creation of Cooper pairs. Similarly enhanced antiferromagnetic spin fluctuations near $T_c$ have also been found in other members of the "122" family, such as isovalently doped $BaFe_2(As_{1-x}P_x)_2$ [252] and representatives of the "11" family. An example of the latter is the stoichiometric superconductor FeSe, whose $T_c$ and in the same time the observed increase of $(T_1T)^{-1}$ are enhanced under pressure [253].

However, recently published $^{77}$Se NMR measurements on $K_xFe_{2-y}Se_2$ reported the same temperature dependence of $(T_1T)^{-1}$ as observed in $LaO_{0.9}F_{0.1}FeAs$: $(T_1T)^{-1}$ decreases from room temperature down to $T_c \approx 33$ K without any hint for antiferromagnetic spin fluctuations [254]. This behavior was explained with the absence of hole bands in the Fermi surface of $K_xFe_{2-y}Se_2$ and the resulting lack of nesting in this compound.

The very different relationship between spin fluctuations and $T_c$ in the aforementioned materials leaves the question open upon their role in the interplay between magnetism and superconductivity in the pnictides in general. The different behavior observed for $LaO_{1-x}F_xFeAs$ on one side and $Ba(Fe_{1-x}Co_x)_2As_2$ and $BaFe_2(As_{1-x}P_x)_2$ on the other side might be related to the different phase diagrams of both pnictide families. These are plotted in Fig. 6.12. The phase diagram of $LaO_{1-x}F_xFeAs$ in Fig. 6.12(a) shows a sharp distinction between the magnetically-ordered phase and the homogeneous superconducting regime with a first order boundary, while there exists a coexistence region of SDW magnetism and superconductivity in the phase diagrams of $Ba(Fe_{1-x}Co_x)_2As_2$ and $BaFe_2(As_{1-x}P_x)_2$ [230, 252, 255–258]. As an example Fig. 6.12(b) reproduces the phase diagram of $Ba(Fe_{1-x}Co_x)_2As_2$ from [230]. The authors of [252] suggest the existence of an antiferromagnetic quantum critical point (QCP) in the phase diagram of $BaFe_2(As_{1-x}P_x)_2$ near the phase boundary of the antiferromagnetic phase close to the doping concentration where $T_c$ is maximal.

One can further check the importance of antiferromagnetic spin fluctuations by plotting the normalized spin-lattice relaxation rate $(T_1T)^{-1}/(T_1T)_0^{-1}$ versus the normalized temperature $T/T_c$. This has been done for several unconventional superconductors such as the heavy fermion system $CeCoIn_5$, the cuprate $YBa_2Cu_3O_7$, and $PuCoGa_5$ by Curro *et al.* [95]. The corresponding plot has been reproduced from [95] in Fig. 6.13(a). It also contains the normalized $(T_1T)^{-1}/(T_1T)_0^{-1}$ of the *s*-wave superconductors Aluminum (only for $T \leq T_C$) and $MgB_2$. $MgB_2$ exhibits a constant $(T_1T)^{-1}$ in the normal state (Fermi liquid behavior), as expected for a conventional BCS superconductor [249]. Conversely, $(T_1T)^{-1}$ of the unconventional superconductors increases Curie-Weiss-likely with decreas-

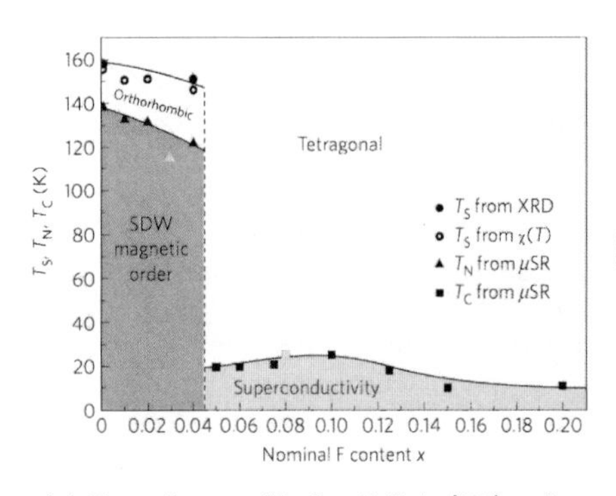


(a) Phase diagram of $LaO_{1-x}F_xFeAs$ [135] as discussed already in Section 4.2.

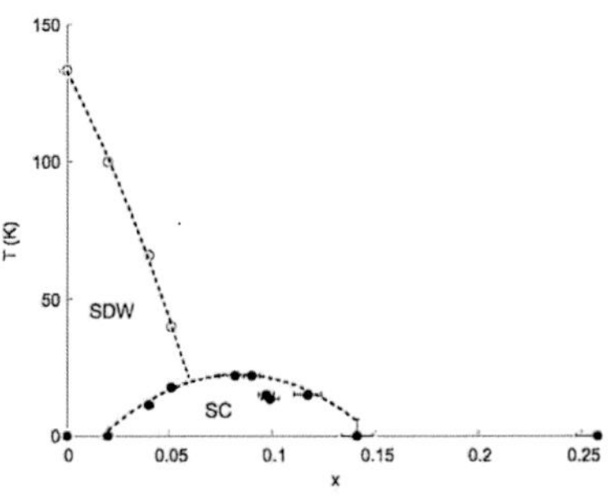


(b) Phase diagram of $Ba(Fe_{1-x}Co_x)_2As_2$ as plotted in [230] (open symbols denote $T_N$, closed symbols denote $T_c$). Similar phase diagrams have been reported in [255–258].

**Figure 6.12:** Comparison of the phase diagrams of $LaO_{1-x}F_xFeAs$ and $Ba(Fe_{1-x}Co_x)_2As_2$. Note that the magnetic and structural phase transitions occur at approximately the same temperature $T_N \approx T_s$ in $Ba(Fe_{1-x}Co_x)_2As_2$.

ing temperature. A scaling of the enhancement of $(T_1T)^{-1}$ in the normal state is found, which suggests that antiferromagnetic spin fluctuations act as an important glue for superconductivity. Fig. 6.13(b) shows that the same scaling works nicely for $Ba(Fe_{1-x}Co_x)_2As_2$ samples with $x \leq 0.12$, FeSe and underdoped $LaO_{0.95}F_{0.05}FeAs$. In contrast, the normalized $(T_1T)^{-1}$ of $LaO_{0.9}F_{0.1}FeAs$ neither falls on this common scaling curve nor shows a constant Fermi-liquid behavior. The superconducting pairing interaction in $LaO_{1-x}F_xFeAs$ is thus unlikely to be mediated by antiferromagnetic spin fluctuations. Note that one has to be cautious when performing such a comparison, since Fig. 6.13(a) only shows unconventional *d*-wave superconductors with nodes in the superconducting gap function (besides $MgB_2$). As will be shown in Chapter 7, the gap symmetry in the pnictides is rather found to be of unconventional $s_\pm$ or $s_{++}$-wave symmetry. Anyhow, already the overall temperature dependence of $(T_1T)^{-1}$ in $LaO_{0.9}F_{0.1}FeAs$ alone (see Fig. 6.9) questions the importance of antiferromagnetic spin fluctuations for superconductivity in this compound.

Concering the relation between optimal superconductivity with the highest $T_c$ and the absence of antiferromagnetic spin fluctuations (on the time scale of NMR) please note the following: $LaO_{0.9}F_{0.1}FeAs$ is frequently denominated as the optimally-doped sample, because it displays the highest $T_c$ within the fluorine doping series of $LaO_{1-x}F_xFeAs$. However, as will be shown in Section 7.2, a sample with an artificially enhanced impurity concentration in the form of arsenic vacancies, $LaO_{0.9}F_{0.1}FeAs_{1-\delta}$, shows an enhanced $T_c = 28.5\,\mathrm{K}$ and in the same time a slightly enhanced $(T_1T)^{-1}$ in the normal state (see Fig. 7.7). The relevance of spin fluctuations for the occurrence of superconductivity in $LaO_{1-x}F_xFeAs$ can thus not be ruled out completely by the absence of such fluctuations in "optimally-doped" $LaO_{0.9}F_{0.1}FeAs$.

#### 6.2.3.2 Quantitative Discussion

To further discuss the strength of antiferromagnetic spin fluctuations in undoped and underdoped $LaO_{1-x}F_xFeAs$, the spin-lattice relaxation rates will now be discussed in a more quantitative fashion. Since a similar set of doping levels of $LaO_{1-x}F_xFeAs$ have been published and analyzed in [168], the first attempt to describe our data theoretically follows their suggestions. Fig. 6.14(a) shows $T_1^{-1}$ of undoped LaOFeAs. From room temperature down to the temperature of the structural phase transition, $T_s = 156\,\mathrm{K}$, $T_1^{-1}$ only slightly increases, while it diverges very fast from $T_s$ down to the magnetic phase transition at $T_N = 138\,\mathrm{K}$. The onset of the divergence of $T_1^{-1}$ just below $T_s$ points towards a close connection between the structural and the magnetic phase transition. Following the same analysis as given in [168], the data were fit with Moriya's self consistent renormalization theory (SCR) for weak itinerant antiferromagnets [259, 260]:

$$\frac{1}{T_1} = \begin{cases} c_1T + c_2T/\sqrt{T - T_N} & T > T_N \\ c_3T/M(T) & T < T_N\,, \end{cases} \tag{6.9}$$

where $c_1, c_2$ and $c_3$ are constants and $M(T)$ is the antiferromagnetic order parameter. Since only one data point was collected in the magnetically ordered state $T < T_N$, only the $T > T_N$ region was fitted. The resulting fit is shown in Fig. 6.14(a) as a solid line. While it nicely reproduces the magnetic ordering temperature $T_N = (137 \pm 1)\,\mathrm{K}$ and the

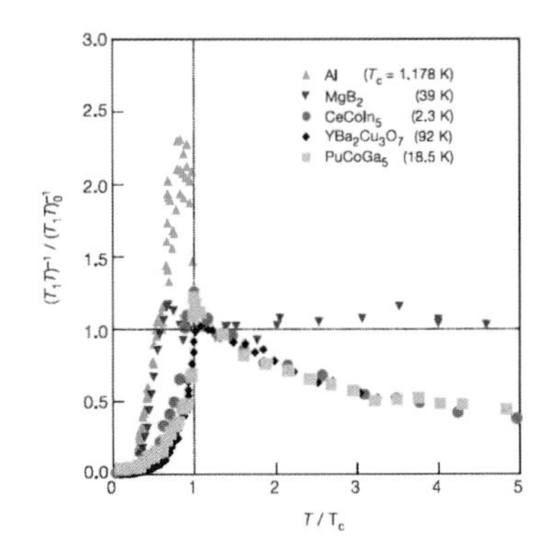


(a) Normalized $(T_1T)^{-1}$ for the unconventional d-wave superconductors $PuCoGa_5$ (yellow squares), $YBa_2Cu_3O_7$ (black rhombi), $CeCoIn_5$ (blue dots), and the s-wave superconductors Al (green triangles) and $MgB_2$ (red triangles). Figure reproduced from [95].

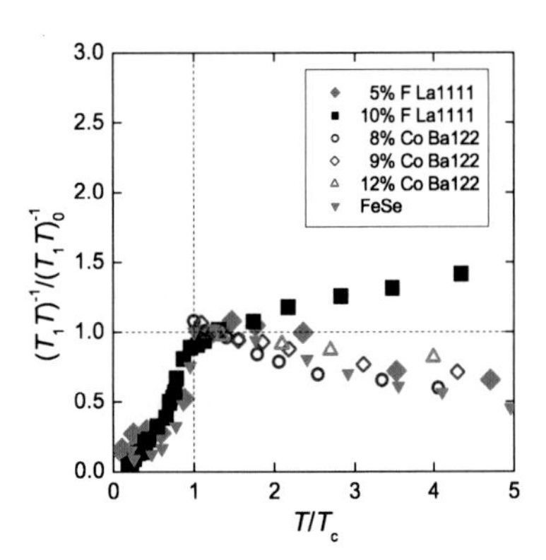


(b) Normalized $(T_1T)^{-1}$ for $LaO_{0.95}F_{0.05}FeAs$ (blue rhombi) and $LaO_{0.9}F_{0.1}FeAs$ (black squares) as well as $Ba(Fe_{1-x}Co_x)_2As_2$ with $x = 0.08$ (open blue circles), $x = 0.09$ (open red rhombi), $x = 0.12$ (open orange triangles) and FeSe (green triangles). Data for $Ba(Fe_{1-x}Co_x)_2As_2$ and FeSe are reproduced from [230] and [253], respectively.

**Figure 6.13:** Normalized $(T_1T)^{-1}$ versus normalized temperature $T/T_c$ for (a) different known superconductors [95] in comparison to (b) the pnictides. The normalization constant $(T_1T)_0^{-1}$ is given by the value of $(T_1T)^{-1}$ at $T = 1.25T_c$.

general trend, it does not capture the differences in slopes above and sligthly below $T_s$. These tiny deviations suggest that the structural distortion indeed has some influence on the magnetic fluctuations preceding the ordered state and should be taken into account for a complete theoretical description of the data. Also multiband effects could play a role.

As already discussed, the underdoped samples with $x = 0.05$ and $x = 0.075$ show a Curie-Weiss-like increase of $(T_1T)^{-1}$ with decreasing temperature. The very first naive approach consists therefore in fitting the data with a simple Curie-Weiss law [168]:

$$\frac{1}{T_1T} = \frac{C}{T+\theta}, \tag{6.10}$$

where $C$ is a constant and $\theta$ the Curie-Weiss temperature. The resulting fits are shown in Fig. 6.14(b). Table 6.2 summarizes the obtained fit parameters for the two samples presented in Fig. 6.14(b) as well as for $LaO_{0.96}F_{0.04}FeAs$ which was presented and analyzed in [168]. Very similar values of $C$ were obtained for $x = 0.05$ and $x = 0.075$. They compare well to $C = 44\text{s}^{-1}$ for $x = 0.04$ [168]. The Curie-Weiss temperature $\theta$ seems to increase with increasing doping, as observed also in $BaFe_2(As_{1-x}P_x)_2$ [252] and $Ba(Fe_{1-x}Co_x)_2As_2$ [230]. However, in contrast to these "122" systems, where a crossover from $\theta < 0$ to $\theta > 0$ has been found (for superconducting samples) and $\theta = 0$ has been discussed within the framework of an antiferromagnetic QCP near the phase boundary of

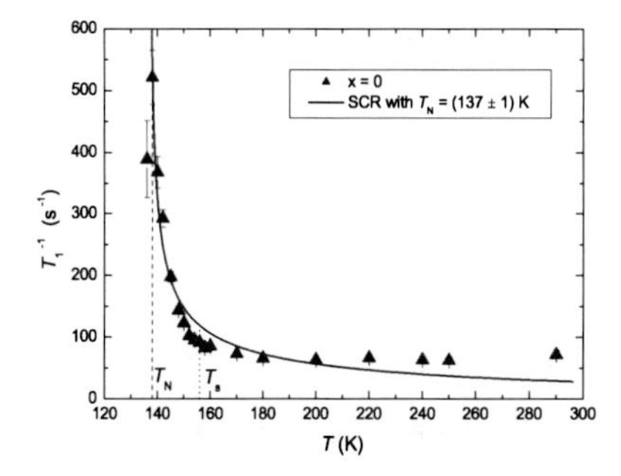


(a) $T_1^{-1}$ of LaOFeAs (triangles) and the corresponding SCR fit for weak itinerant antiferromagnets according to Eq. (6.9) (solid line). The dotted and dashed vertical lines denote $T_s$ and $T_N$, respectively.

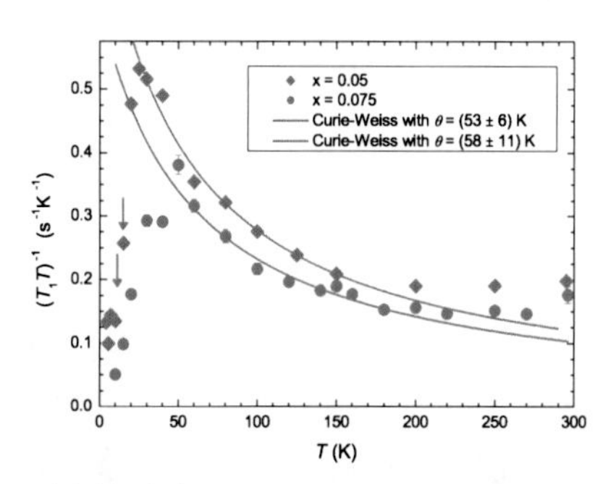


(b) $(T_1T)^{-1}$ of $LaO_{1-x}F_xFeAs$ with $x = 0.05$ (blue squares) and $x = 0.075$ (green dots) and the corresponding Curie-Weiss fits (solid lines) according to Eq. (6.10). Arrows denote $T_c$.

**Figure 6.14:** SCR fits for undoped LaOFeAs (a) and Curie-Weiss fits for underdoped $LaO_{1-x}F_xFeAs$ with $x = 0.05$ and $x = 0.075$ (b).

the antiferromagnetically ordered SDW state, the Curie-Weiss temperature of superconducting $LaO_{1-x}F_xFeAs$ is always positive. This might be due to the difference in phase diagrams as already discussed in Section 6.2.3.1 (see Fig. 6.12). Note that an increase of $\theta$ with increasing Sr concentration and a nearly vanishing $\theta$ near the magnetic phase boundary has also been observed in the cuprate $La_{2-x}Sr_xCuO_4$ [261].

While the simple Curie-Weiss fit presented in Fig. 6.14(b) reflects well the overall trend of an increasing $(T_1T)^{-1}$ with decreasing temperature, it fails in two major points. First of all, it is unable to describe the data at high temperature. Second, it misses a very peculiar feature of $(T_1T)^{-1}$ just above $T_c$. By closely examining the data, one observes that $(T_1T)^{-1}$ actually shows a maximum clearly above $T_c$ and starts to decrease already above the actual onset of superconductivity at $T_c$. Such a peak was also observed in $LaO_{0.96}F_{0.04}FeAs$ and interpreted as a weak magnetic ordering [168]. Also NMR measurements on superconducting FeSe, at ambient conditions as well as under pressure, reported the occurrence of such a peak in $(T_1T)^{-1}$ above the onset of superconductivity [253]. They related it to enhanced spin fluctuations which might cause a glassy spin freezing before

| $x$ | fitting range | $C$ (s$^{-1}$) | $\theta$ (K) |
|---|---|---|---|
| 0.04 [168] | 30 K - 250 K | 44 | 10.2 |
| 0.05 | 25 K - 150 K | $43 \pm 3$ | $53 \pm 6$ |
| 0.075 | 60 K - 180 K | $37 \pm 3$ | $58 \pm 11$ |

**Table 6.2:** Curie-Weiss fitting parameters for underdoped $LaO_{1-x}F_xFeAs$. The first row ($x = 0.04$) refers to a fit on $(T_1T)^{-1}$ of $LaO_{0.96}F_{0.04}FeAs$ reported in [168].

the onset of bulk superconductivity. Such a peak in $(T_1T)^{-1}$ cannot be described within a simple Curie-Weiss law.

To better describe the peak slightly above $T_c$ and the linear decrease at high temperature, the following fitting function was used:

$$\frac{1}{T_1T} = a + bT + \left(\frac{1}{T}\right)\frac{c\tau_c}{1+\tau_c^2\omega_L^2}\,. \tag{6.11}$$

$a$ and $b$ are constants. The last part of this equation is the BPP model introduced in Eq. (2.51). It describes the behavior of $T_1^{-1}$ upon the influence of fluctuating magnetic fields $h_\perp$. The prefactor $c = \gamma^2 h_\perp^2$ contains the fluctuating magnetic field $h_\perp$ perpendicular to the externally applied magnetic field direction. At the temperature where $T_1^{-1}$ is maximal, the effective correlation time $\tau_c$ of the spin fluctuations is just the inverse Larmor frequency of the nuclear spins: $\tau_c = \omega_L^{-1}$. The correlation time of the spin fluctuations is temperature-dependent. For a glassy spin freezing, this dependence can be described by an activated behavior [65, 66, 224] (see also Eq. (2.52)):

$$\tau_c = \tau_0 \exp(E_a/k_BT)\,. \tag{6.12}$$

In this equation $\tau_0$ is the correlation time at high temperature and $E_a$ is the activation energy of the spin fluctuations. A fit to the data considering only the BPP model [last part of Eq. (6.11)] did not yield satisfying results. A simple addition of a constant value $a$ as done by other groups for example in their Curie-Weiss fits [252], also did not describe the data well. To account for the high temperature behavior of $(T_1T)^{-1}$, a linear temperature dependence had to be added, resulting in the final fit equation (6.11). The inset of Fig. 6.15 shows $(T_1T)^{-1}$ of the samples with $x = 0.05$ and $x = 0.075$ and the corresponding fits according to Eq. (6.11) with free fit parameters. The fit and the data agree within error bars in the whole paramagnetic temperature range.

As discussed in the qualitative description of the spin-lattice relaxation rate data, spin fluctuations seem to be totally suppressed in the optimally-doped sample. $(T_1T)^{-1}$ of the optimally-doped $LaO_{0.9}F_{0.1}FeAs$ follows a linear temperature dependence in the whole paramagnetic temperature range. A linear fit of the form $(T_1T)^{-1} = a + bT$ with $a = (0.052 \pm 0.002)\,\mathrm{s^{-1}K^{-1}}$ and $b = (3.2 \pm 0.1) \times 10^{-4}\,\mathrm{s^{-1}K^{-2}}$ is shown in Fig. 6.15 for $LaO_{0.9}F_{0.1}FeAs$. It describes the temperature dependence of $(T_1T)^{-1}$ in the paramagnetic state perfectly. Let's assume, that the linear temperature dependence of $(T_1T)^{-1}$ at high temperature in the underdoped samples is of the same origin as the linear temperature dependence of $(T_1T)^{-1}$ of the optimally-doped sample and is not related to the antiferromagnetic fluctuations which lead to an increase of $(T_1T)^{-1}$ in the underdoped samples at "intermediate" temperatures. The linear part of Eq. (6.11) should then be independent of doping and follow the same behavior as the optimally-doped $LaO_{0.9}F_{0.1}FeAs$. The $(T_1T)^{-1}$ data of the underdoped samples were therefore fitted with Eq. (6.11), but with $a$ and $b$ fixed to the values obtained by the linear fit to $(T_1T)^{-1}$ of $LaO_{0.9}F_{0.1}FeAs$. The results shown in Fig. 6.15 are astonishingly good. As in the case of free parameters, the data and the fit agree nicely with each other in the whole considered temperature range and for both doping levels. Table 6.3 summarizes the obtained fitting parameters. Similar values have been obtained by fitting the data with free fitting parameters.

The last row of Table 6.3 contains the value of the fluctuating field, obtained from the fitting parameter $c = \gamma^2 h_\perp^2$. Since such an analysis have not been performed on

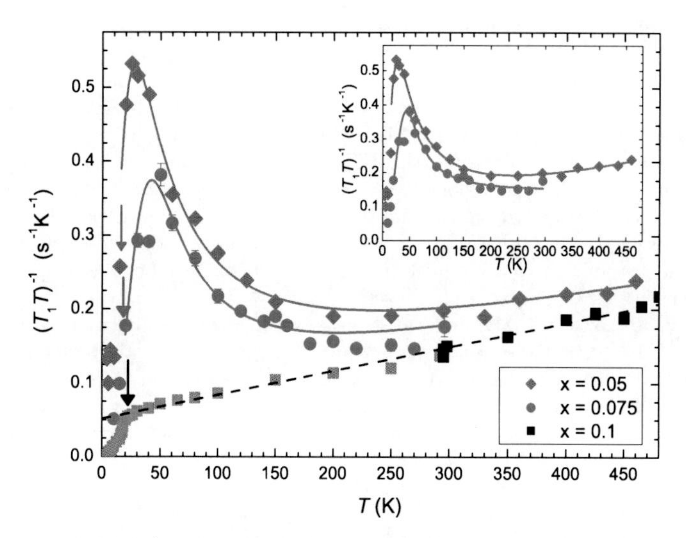


**Figure 6.15:** $(T_1T)^{-1}$ of $LaO_{0.95}F_{0.05}FeAs$ (blue squares), $LaO_{0.925}F_{0.075}FeAs$ (green dots) and $LaO_{0.9}F_{0.1}FeAs$ (black and grey squares). The dashed line is a linear fit $a + bT$ to $(T_1T)^{-1}$ for $LaO_{0.9}F_{0.1}FeAs$. Solid lines are fits according to Eq. (6.11), adopting $a$ and $b$ from the linear fit to $LaO_{0.9}F_{0.1}FeAs$. Arrows denote the corresponding $T_c(H_0)$ ($\approx$ 16 K, 18 K and 22 K for $x$ = 0.05, $x$ = 0.075 and $x$ = 0.1). The inset shows the data for $LaO_{0.95}F_{0.05}FeAs$ and $LaO_{0.925}F_{0.075}FeAs$ and their corresponding BPP fits with totally free parameters.

other pnictides before, the values can not be compared to similar systems. A peak in the spin-lattice relaxation rate has also been observed in the lightly doped cuprates $La_{2-x}Sr_xCuO_4$, $La_{1.8-x}Eu_{0.2}Sr_xCuO_4$ and $La_2Cu_{1-x}Li_xO_4$ in the striped ordered phase, suggesting a glassy, quasi-static spin freezing [65, 66, 262, 263]. This slowing down of spin fluctuations has also been evaluated within the BPP model. The resulting activation energies $E_a$ lie in between 60 K and 120 K and the characteristic correlation time of the spin fluctuations lie in between $10^{-13}$ and $10^{-14}$ seconds. While $E_a$ deduced for the underdoped pnictides is of the same order of magnitude as $E_a$ in these cuprate systems, the characteristic correlation time of $LaO_{1-x}F_xFeAs$ is 3-4 orders of magnitude larger, which means that the fluctuations are much slower $LaO_{1-x}F_xFeAs$ then in the mentioned

| $x$ | $E_a$ (K) | $\tau_0$ ($10^{-10}$ s) | $h_\perp$ (Oe) |
|---|---|---|---|
| 0.05 | $40 \pm 2$ | $18 \pm 1$ | 150 |
| 0.075 | $78 \pm 2$ | $8.5 \pm 0.5$ | 135 |

**Table 6.3:** BPP fitting parameters for underdoped $LaO_{1-x}F_xFeAs$.

cuprate materials. The reason of this discrepancy is unclear. A correlation time in the range of $10^{-10}$ s is really slow in solid state physics.

Electron spin resonance (ESR) measurements on $GdO_{1-x}F_xFeAs$ found signatures for short-range magnetic correlations in underdoped samples of $GdO_{1-x}F_xFeAs$ with nominal fluorine contents of $x = 0.07$ and $x = 0.14$ [143]. These magnetic correlations are quasi-static on the time scale of high frequency ESR, which corresponds to pico seconds. This is still two orders of magnitude shorter than the characteristic correlation time which was needed to describe the NMR data within the BPP model. But it is consistent with the NMR results in the sense that NMR is a "slower" probe than ESR. Spin fluctuations which from the point of view of NMR are freezing, should correspondingly appear also quasi-static when seen through the "glasses" of ESR.

Muon spin relaxation ($\mu$SR) and Mössbauer measurements on underdoped and optimally-doped $LaO_{1-x}F_xFeAs$ could rule out any static SDW magnetism with considerable magnetic moments in these compounds [135, 218]. However, a slight increase of the $\mu$SR $\lambda_{ab}^{-2}$ for the underdoped samples at low temperature, where $\lambda_{ab}$ is the in-plane magnetic penetration depth, and a gradual increase of the linewidth of the Mössbauer spectra for $LaO_{0.95}F_{0.05}FeAs$ below 10 K indicate the presence of dilute magnetic correlations [135, 218]. In $\mu$SR an additional relaxation due to tiny static internal magnetic fields of electronic origin is observed at low temperature for both underdoped samples [135]. The corresponding internal magnetic fields measured by the muon relaxation rate and the muon precession frequency are found to be 20 times smaller in $LaO_{0.95}F_{0.05}FeAs$ than in the magnetically-ordered state of $LaO_{0.96}F_{0.04}FeAs$ [135]. Together with the reported value of the internal field at the muon site in the ordered state of $LaO_{0.96}F_{0.04}FeAs$, which amounts to $H_{int,\mathrm{muon}} = 160\,\mathrm{mT}$ [135], this results in local internal fields of the order of 80 Oe in $LaO_{0.95}F_{0.05}FeAs$, which is of the same order of magnitude as the values deduced from the BPP fit to $(T_1T)^{-1}$ of $LaO_{0.95}F_{0.05}FeAs$. Note that the additional relaxation is only observed in a minor volume fraction of the sample, indicating dilute magnetic clusters and the absence of coherent magnetism [135].

Some important issues should be mentioned at this point:

**1.** The standard use of the BPP model is to describe a peak in the spin-lattice relaxation rate $T_1^{-1}$ itself. Within the analysis presented here, it was used to describe the peak in the spin-lattice relaxation rate divided by temperature, $(T_1T)^{-1}$. On that account, the prefactor $1/T$ has been added in Eq. (6.11). If one would plot the bare spin-lattice relaxation rate, $T_1^{-1}$ versus temperature, no pronounced characteristic BPP peak would be visible, just a weak anomaly. The temperature dependence, especially at high temperatures, is therefore a crucial point. Only after subtracting the linear temperature dependence of $(T_1T)^{-1}$ (which corresponds to $T_1^{-1} \propto T^2$) at high temperatures, a peak in $T_1^{-1}$ gets visible. The data are therefore determined by two different effects: a non-Fermi-liquid-like linear temperature dependence of $(T_1T)^{-1}$ at high temperatures, and the slowing down of antiferromagnetic spin fluctuations at low temperatures. The superposition of these two effects complicate detailed quantitative analyses. The values given above should therefore be taken with care.

**2.** A similar approach to describe $(T_1T)^{-1}$ of the underdoped samples with a high temperature linear behavior and a low temperature Curie-Weiss increase was also tried. It did not give any satisfactory results. This was also the case for a Curie-Weiss behavior extended by an additive constant offset: $1/T_1T = \text{const.} + C/(T+\theta)$, which was used by other authors [252]. Similarly, the SCR theory applied to the data of LaOFeAs did not yield better results when considering an additional constant or an additional linear temperature behavior.

**3.** For the $LaO_{0.96}F_{0.04}FeAs$ sample presented in [168], a simple Curie-Weiss fit gave satisfactory results at intermediate temperature. Note however that in this case only data up to 250 K were shown. A deviation from the Curie-Weiss fit is expected to occur also in this sample at higher temperatures. Furthermore, $(T_1T)^{-1}$ already decreases above $T_c$ also for this doping concentration, showing a similar maximum as has been observed in $LaO_{0.95}F_{0.05}FeAs$ and $LaO_{0.925}F_{0.075}FeAs$, which cannot be described by a simple Curie-Weiss law.

### 6.2.4 Summary

Spin dynamics of $LaO_{1-x}F_xFeAs$ have been investigated by $^{75}$As $(T_1T)^{-1}$ for a whole set of doping levels ranging from undoped LaOFeAs up to optimally-doped $LaO_{0.9}F_{0.1}FeAs$. For the latter, $(T_1T)^{-1}$ was also measured at the $^{139}$La site and compared to $^{57}$Fe and $^{19}$F $(T_1T)^{-1}$ data.

The undoped compound shows a strong increase of $^{75}(T_1T)^{-1}$ towards the magnetic ordering temperature $T_N$, which can be fairly well described by Moriya's SCR theory for weak itinerant antiferromagnets. A relation between the structural phase transition and the magnetic phase transition is assumed, since the divergence of $(T_1T)^{-1}$ starts around $T_s$.

The underdoped samples with $x = 0.05$ and $x = 0.075$, where the SDW state is suppressed and superconductivity is observed at low temperature, still show a substantial increase of $(T_1T)^{-1}$ with decreasing temperature at intermediate temperatures and a linear temperature dependence of $^{75}(T_1T)^{-1}$ at higher temperatures. While the first feature reflects remnant AFM fluctuations, the linear temperature dependence resembles the non Fermi-liquid behavior of the optimally-doped sample. The data for $x = 0.05$ and $x = 0.075$ could be nicely fit with a combination of a linear term and a BPP model of fluctuating magnetic fields over the whole paramagnetic temperature range.

$^{75}(T_1T)^{-1}$ of optimally-doped $LaO_{0.9}F_{0.1}FeAs$ does not show any hint for magnetic correlations any more. It decreases linearly with temperature from 480 K down to the onset of superconductivity. This behavior is clearly non Fermi-liquid-like. Pseudogap fits as suggested and carried by other groups did not give satisfactory results. A scaling of $(T_1T)^{-1}$ data of different nuclei measured in optimally-doped $LaO_{0.9}F_{0.1}FeAs$ showed, that spin fluctuations are suppressed simultaneously across the whole $\vec{q}$-space and the spin-lattice relaxation rate is mostly dominated by simple quasi particle scattering.

The overall doping evolution of the temperature-dependent spin-lattice relaxation rate suggests that antiferromagnetic spin fluctuations are still present in underdoped superconducting samples. $T_c$ however only reaches its maximum value as soon as these fluctuations are suppressed (at least at the time scale of NMR). These results stand in contrast to NMR measurements on representatives of the "122" family such as $Ba(Fe_{1-x}Co_x)_2As_2$ and $BaFe_2(As_{1-x}P_x)_2$, where antiferromagnetic fluctuations are still well pronounced in samples with the highest $T_c$. These differences were discussed based on the different phase diagrams of both pnictide families.

## 6.3 Korringa Relation

Now that both the static and dynamic NMR properties of $LaO_{1-x}F_xFeAs$ have been discussed separately in the former sections, they can be analysed with regard to their relation to each other. According to Section 2.3.2.4, one can describe a system within a Fermi liquid picture, if the ratio $K_s^2 T_1 T$, the so-called Korringa ratio, is constant. Such a constant behavior has been observed for the $^{75}As$ nuclei in $LaO_{0.9}F_{0.1}FeAs$ in the paramagnetic temperature region up to $T = 300\,K$ [79]. It suggests that either $^{75}As$ is insensitive to antiferromagnetic spin fluctuations or antiferromagnetic spin fluctuations have disappeared at this doping level. The first assumption can be ruled out by $(T_1T)^{-1}$ measurements on the $^{75}As$ nuclei in underdoped samples, which showed a strong upturn upon lowering temperature and thus prove the sensitivity of $^{75}As$ for antiferromagnetic fluctuations. The observation of the Korringa ratio in $LaO_{0.9}F_{0.1}FeAs$ therefore indicates, that at this

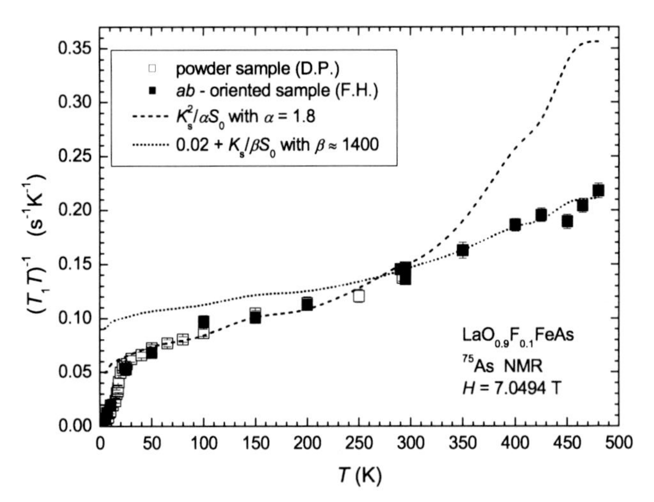


**Figure 6.16:** $(T_1T)^{-1}$ vs temperature measured at $H \parallel ab$ for a powder (open squares) and an *ab*-oriented sample (filled squares) of $LaO_{0.9}F_{0.1}FeAs$. Dashed and dotted lines indicate $(T_1T)^{-1} \propto K_s^2$ and $(T_1T)^{-1} \propto K_s$, respectively. The powder data and the Korringa ratio $(T_1T)^{-1} \propto K_s^2$ up to $T = 300\,K$ are reproduced from [79].

(optimal) doping level, antiferromagnetic spin fluctuations have disappeared completely. The data of [79] (open squares) are reproduced in Fig. 6.16. The corresponding Korringa ratio $(T_1T)^{-1} = K_s^2/\alpha S_0$ with $\alpha = 1.8$ [according to Eq. (2.59)] is sketched as a dashed line. The new data up to $T = 480$ K (filled squares) have been added to Fig. 6.16. Based on these new high temperature data for $(T_1T)^{-1}$ as well as on the new data for $K_s$, the Korringa ratio $(T_1T)^{-1} = K_s^2/1.8S_0$ has been extended up to $T = 480$ K (dashed line). A strong deviation from this Korringa ratio is visible for temperatures $T > 300$ K. In this high temperature regime, the $(T_1T)^{-1}$ data rather follow a linear dependence on the spin shift $K_s$ (dotted line), which in turn does not well describe the data for $T < 300$ K.

In the case of underdoped $LaO_{0.95}F_{0.05}FeAs$, no constant Korringa ratio $K_s^2T_1T$ could be found at any temperature range. But surprisingly, for $T \geq 250$ K the same linear relationship between $(T_1T)^{-1}$ and $K_s$ was found as in $LaO_{0.9}F_{0.1}FeAs$ for $T \geq 300$ K (see Fig. 6.17). It differs only in the absolute value of the $y$-intercept (0.02 versus 0.075). This might be related to the difficulty in determining $K_{ab}$ and therewith $K_{orb}$ in $LaO_{0.95}F_{0.05}FeAs$ due to the complicated $^{75}$As-NQR spectrum (see Section 6.1.2).

For undoped LaOFeAs and $LaO_{0.925}F_{0.075}FeAs$ no comment can be made on the relation between $T_1T$ and $K_s$, since $K_s$ has not been measured in these compounds.

In agreement with the observation of a constant Korringa ratio for $LaO_{0.9}F_{0.1}FeAs$ and its absence in $LaO_{0.95}F_{0.05}FeAs$, no universal relation between $K_s$ and $(T_1T)^{-1}$ was found for the pnictides in general. A linear scaling behavior has also been observed in $LaO_{0.89}F_{0.11}FeAs$ for $T_c < T \leq 300$ K via $^{19}$F-NMR [226]. In $Ba(Fe_{1-x}Co_x)_2As_2$, the

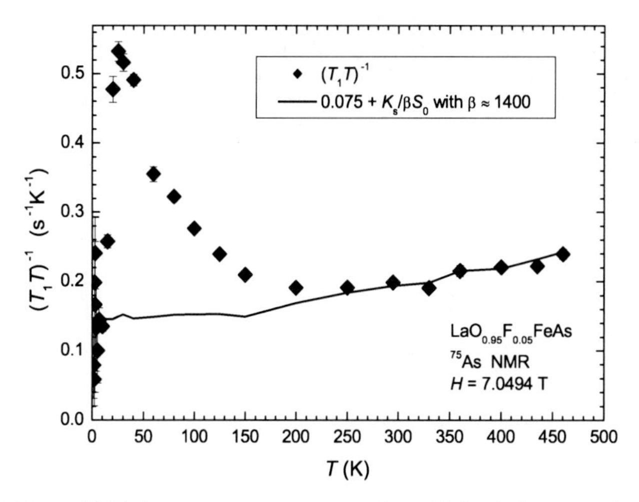


**Figure 6.17:** $(T_1T)^{-1}$ vs temperature measured at $H \parallel ab$ for a powder sample of $LaO_{0.95}F_{0.05}FeAs$ (black rhombi). The solid line indicates $(T_1T)^{-1} \propto K_s$ with the same proportionality constant $\beta$ as in Fig. 6.16 for $LaO_{0.9}F_{0.1}FeAs$. $K_s(T)$ was determined as $K_{ab,mean}(T) - K_{orb}$ with $K_{orb} = 0.01$ (see Section 6.1.2).

Korringa ratio $K_s^2 T_1 T = \text{const.}$ was found only below $T^* = 100\,\text{K}$ in a nearly optimally-doped sample ($x = 0.1$) [166] and in the whole temperature range of $4.2\,K \leq T \leq 295\,\text{K}$ for a heavily overdoped, non-superconducting sample with $x = 0.26$ [230]. While in the case of $Ba(Fe_{0.9}Co_{0.1})_2As_2$ the observation of $K_s^2 T_1 T = \text{const.}$ stands in contrast to the non-Fermi-liquid behavior of the resistivity, which follows a linear temperature dependence [166, 264], the resistivity of the heavily overdoped sample ($x = 0.26$) scales as $\rho \propto T^2$ and thus confirms the Fermi-liquid picture [230]. The authors of [230] suggest that a Fermi-liquid description of the spin excitations in $Ba(Fe_{1-x}Co_x)_2As_2$ is only possible at high doping levels, when the hole Fermi surface at the center of the Brillouin zone is completely filled by electron doping and thus disappears, leaving behind only electron pockets and therewith only intraband electron excitations [230].

The temperature-dependent resistivity $\rho(T)$ of $LaO_{1-x}F_xFeAs$ agrees nicely with the observed Fermi-liquid Korringa ratio for $T_c \leq T \leq 300\,\text{K}$ in $LaO_{0.9}F_{0.1}FeAs$ and the lack of such a behavior in $LaO_{0.95}F_{0.05}FeAs$. Only for high doping levels $0.1 \leq x \leq 0.2$, it shows a quadratic temperature dependence, $\rho(T) \propto T^2$, in a broad temperature range from $T_c$ up to $T \approx 200\,\text{K}$, indicating a Fermi-liquid state [179, 180, 265]. At higher temperatures, it depends only linearly on temperature, $\rho(T) \propto T$ [179, 180]. According to the phase diagram based on the temperature dependence of $\rho(T)$ proposed in [180], a Fermi-liquid state seems to exist only in optimally or overdoped samples at low temperatures.

Correspondingly, NMR measurements on overdoped samples with $x > 0.1$ would be interesting. They should also show a constant Korringa ratio, reflecting the Fermi-liquid nature suggested by resistivity measurements [179, 180]. The proposed linear relationship between $(T_1T)^{-1}$ and $K_s$ in $LaO_{0.89}F_{0.11}FeAs$ [226] stands in contrast to a Fermi liquid behavior for doping levels $x \geq 0.1$, which was suggested by resistivity measurements. However, a careful revision of the data of [226] illustrates that a quadratic dependence of $(T_1T)^{-1}$ on $K_s$, indicative of a Fermi-liquid state, cannot be excluded (see Fig. 6.18). Note that the choice between a linear or quadratic dependence of $(T_1T)^{-1}$ on $K_s$ depends delicately on the temperature-independent orbital shift $K_{orb}$ [23, 266].

What can be learned about the correlations in the Fermi-liquid phase in $LaO_{0.9}F_{0.1}FeAs$ from the Korringa ratio? It suggests that antiferromagnetic fluctuations of the iron $3d$ moments at $\vec{Q}_{af} = (\pi, \pi)$, which play a significant role at lower doping levels (leading to the upturn in $(T_1T)^{-1}$ in $LaO_{0.95}F_{0.05}FeAs$ and $LaO_{0.925}F_{0.075}FeAs$), do not contribute to the spin-lattice relaxation rate of $LaO_{0.9}F_{0.1}FeAs$. The absolute value of $\alpha = 1.8 \approx \alpha_0 = 1$ suggests the absence of any strong magnetic correlations in the system (see Section 2.3.2.4).

Concerning the high temperature region where a robust linear relationship between $(T_1T)^{-1}$ and $K_s$ has been found for both $LaO_{0.95}F_{0.05}FeAs$ and $LaO_{0.9}F_{0.1}FeAs$, no detailed analysis of this non-Fermi-liquid behavior can be given. It is in agreement with resistivity measurements, which also show a non-Fermi-liquid $\rho(T) \propto T$ dependence.

It is tempting to ascribe this linear temperature dependence of $\rho(T)$ to conventional phonon scattering, as it is expected in a normal metal at high temperatures ($T > \Theta_D$). However, the Debye temperature of iron amounts to $\Theta_D = 420\,\text{K}$ [215], which is much higher than the onset of the $\rho \propto T$ behavior. Other unknown scattering centers, whose density of states also depends linearly on temperature might thus be responsible for $\rho(T) \propto T$. This might indicate another electronic phase at high temperature. Some references have assigned the linear temperature dependence of the resistivity as a signature of spin fluctuations reminiscent of the SDW ordered state [267, 268]. This explanation is questionable since NMR measurements do not show any hint for remnant spin fluctuations in the temperature and doping regimes, where a linearly temperature-dependent resistivity is observed.

From the point of view of NMR, two different cases are known which also showed a $T_1TK_s$ = const. behavior. The first one are metallic compounds such as $YCO_2$ [269] and $TiBe_2$ [270], where the observed linearity between $(T_1T)^{-1}$ and $K_s$ is in good accord with Moriya's SCR theory for nearly or weakly ferromagnetic metals. This possibility can be excluded for the pnictides presented here due to their proximity to an anitferromagnetic instability. The second class of materials where $T_1TK_s$ = const. was observed are cuprates, especially for $^{17}O$ NMR in $YBa_2Cu_4O_8$ and $YBa_2Cu_3O_{6+x}$ [31, 266, 271–274]. While the real underlying physical origin is still unclear [274], the authors of [31, 266, 271–274] suggested the following interpretation: According to Eq. (6.6), $(T_1T)^{-1}$ is proportional to the dynamic susceptibility divided by the Larmor frequency, summed up over the whole $\vec{q}$-space:

$$\frac{1}{T_1T} \propto \gamma^2 \sum_{\vec{q}} |A_\perp(\vec{q})|^2 \frac{\chi''_\perp(\vec{q}, \omega_L)}{\omega_L}. \tag{6.13}$$

They now assume, that $\chi''_\perp(\vec{q}, \omega_L)/\omega_L$ has a Lorentzian frequency dependence with width $\Gamma_q$ in the paramagnetic state and substitute:

$$\frac{\chi''_\perp(\vec{q}, \omega_L)}{\omega_L} = \frac{\pi\chi(\vec{q})}{\Gamma_q}, \tag{6.14}$$

where $\chi(\vec{q})$ is the $\vec{q}$-dependent static susceptibility. Now they apply the MMP model [50], which phenomenologically devides the contributions to the static $\vec{q}$-dependent susceptibility into two parts, mainly a Fermi-liquid-like $\vec{q}$-independent part from the whole $\vec{q}$-space,

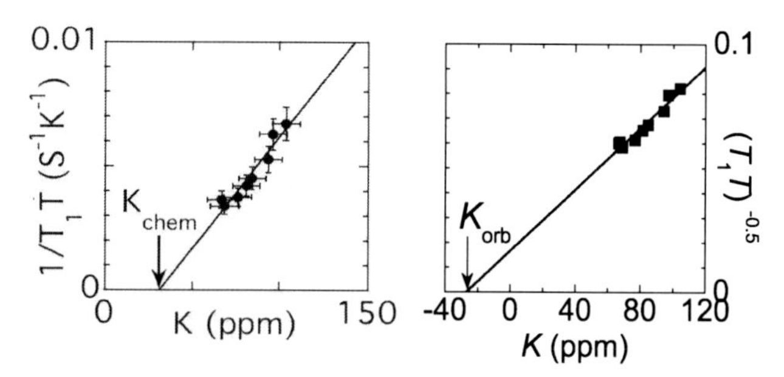


**Figure 6.18:** $(T_1T)^{-1}$ vs $K$ for $LaO_{0.89}F_{0.11}FeAs$. The left panel, which shows $(T_1T)^{-1} \propto K$ has been taken from [226]. For the right panel, the data have been digitized and analyzed with $(T_1T)^{-0.5} \propto K$, reflecting the Korringa relation, which can also describe the data. Note that in both plots $K_{orb} = K_{chem}$ has not been subtracted from $K(T) = K_s(T) + K_{orb}$. It amounts to $K_{orb}$ = 30 ppm in the case of a linear scaling (left panel) and $K_{orb}$ = −27 ppm in the case of a Korringa behavior (right panel).

$\chi(0)$ and another part $\chi_{\rm afm}(\vec{q})$, which expresses the strong antiferromagnetic fluctuations at (or near) $(\pi, \pi)$:

$$\frac{\chi''_\perp(\vec{q}, \omega_L)}{\omega_L} = \frac{\pi\chi(\vec{q})}{\Gamma_q} = \frac{\pi\chi(0)}{\Gamma_0} + \frac{\pi\chi_{\rm afm}(\vec{q})}{\Gamma_{{\rm afm},q}} . \quad (6.15)$$

Since the hyperfine form factor of $^{17}$O nuclei in the cuprates is almost zero at the antiferromagnetic wave vector, $^{17}$O NMR is only sensitive to probe the first part of the MMP model:

$$\frac{1}{T_1T} \propto \frac{\pi\chi(0)}{\Gamma_0} . \quad (6.16)$$

The experimental observation of

$$^{17}\left(\frac{1}{T_1T}\right) \propto K_s \propto \chi_s \quad (6.17)$$

in $YBa_2Cu_4O_8$ and $YBa_2Cu_3O_{6+x}$ then suggests that the energy width of the spin excitations $\Gamma_0$ is temperature-independent, although the spin susceptibility itself shows a strong temperature dependence [31, 266, 271–274].

It is questionable whether a similar scenario might be applicable to the case of pnictides. However, the multiband character of the electronic band structure might allow to play around with different contributions to the dynamic susceptibility from different bands.

## 6.3.1 Summary

This Section discussed the Korringa ratio in $LaO_{0.9}F_{0.1}FeAs$ and $LaO_{0.95}F_{0.05}FeAs$ based on $^{75}$As NMR measurements. A real Korringa ratio $T_1TK_s^2 = $ const. was only observed in optimally-doped $LaO_{0.9}F_{0.1}FeAs$ for $T_c < T < 300$ K. This is in good agreement with resistivity measurements, which reported a Fermi-liquid-like $\rho \propto T^2$ for $T_c < T \leq 300$ K [180]. A linear relation, $T_1TK_s = $ const., was found for $T > 300$ K in the case of $LaO_{0.9}F_{0.1}FeAs$ and for $T \geq 250$ K in the case of $LaO_{0.95}F_{0.05}FeAs$. In both cases the slope was the same, suggesting an intrinsic high temperature electronic phase which differs from the one at low temperature and persists at a broad doping range, independent of the very different spin dynamics in $LaO_{0.95}F_{0.05}FeAs$ and $LaO_{0.9}F_{0.1}FeAs$ at low temperatures. The observation of such a linear relation between $(T_1T)^{-1}$ and $K_s$ goes along with a linear temperature dependence of the resistivity [180]. The physical origin of this behavior is still unclear. A rough explanation was tried using the MMP model which was successfully applied to the cuprates. Since cuprates and iron-based superconductors differ strongly in their basic properties (charge transfer insulators vs itinerant metals, single band vs multiband, ...) this might however not be appropriate. Measurements on the overdoped side of $LaO_{1-x}F_xFeAs$ are still missing. They would be interesting since resistivity measurements on overdoped $LaO_{1-x}F_xFeAs$ [180] and NMR measurements on heavily overdoped, non-superconducting $Ba(Fe_{0.74}Co_{0.26})_2As_2$ [230] suggest a Fermi-liquid-like behavior at high doping levels.

# 7 NMR and NQR on $LaO_{1-x}F_xFeAs$ in the Superconducting State

In the following, NMR and NQR measurements in the superconducting state of underdoped and optimally-doped $LaO_{1-x}F_xFeAs$ with $x = 0.05$, $x = 0.075$ and $x = 0.1$ will be shown and discussed within possible gap symmetries. The presentation of the data is focussed on dynamic properties. Spin-lattice relaxation rate data for $x = 0.05$, $x = 0.075$ and $x = 0.1$ will be shown and analysed in Section 7.1. Section 7.2 reports the special case of $LaO_{0.9}F_{0.1}FeAs_{1-\delta}$, an optimally-doped sample with additional impurities (arsenic vacancies), exhibiting slightly enhanced superconducting properties and an even sharper decrease of $T_1^{-1}$ than optimally-doped $LaO_{0.9}F_{0.1}FeAs$.

NMR Knight shift measurements in the superconducting state of $LaO_{0.9}F_{0.1}FeAs$, $LaO_{0.89}F_{0.11}FeAs$, oxygen-deficient $LaO_{0.7}FeAs$ and $PrO_{0.89}F_{0.11}FeAs$ reported a decreasing Knight shift in the superconducting state, indicating the robustness of spin-singlet superconductivity in the "1111" family [79, 147, 162, 164, 167, 275, 276]. The same behavior holds true for the "122" family [166, 170, 171, 229, 230, 277, 278]. The spin-lattice relaxation rate data presented in the following will therefore be discussed considering the theoretically possible symmetries of the superconducting order parameter for singlet superconductors, namely *s*- and *d*-wave symmetries.

## 7.1 Spin-Lattice Relaxation Rate for $x \geq 0.05$

$^{75}$As NMR $T_1^{-1}$ measurements were performed for $LaO_{1-x}F_xFeAs$ samples with $x = 0.05$, $x = 0.075$ (whole displayed temperature range) and $x = 0.1$ (low temperatures, $T < 10\,K$). They are assembled in Fig. 7.1 together with already published data for $LaO_{0.9}F_{0.1}FeAs$ [79]. Additionally, data for superconducting $LaO_{0.96}F_{0.04}FeAs$ are reproduced from Ref. [168] in Fig. 7.1[1]. For all doping levels, $T_1^{-1}$ decreases rapidly without any signature of a Hebel-Slichter coherence peak below $T_c$, following a $T^3$ dependence down to $T \approx 0.3T_c$. This temperature dependence can be interpreted as an indication for the existence of line nodes in the superconducting gap function (see Chapter 3). Similar $T^3$ dependencies and interpretations were reported in several other early NMR publications [147–150, 277].

The new data points for $LaO_{0.9}F_{0.1}FeAs$ at low temperature (black squares) show a clear deviation from the $T^3$ behavior for $T < 4.2\,K$. Such a low temperature feature with a nearly linear slope below $T \approx 0.2T_c$ was also observed in $LaO_{0.95}F_{0.05}FeAs$ (see Fig. 7.1), $LaO_{0.9}F_{0.1}FeAs_{1-\delta}$ (see Section 7.2, [279]) and $BaFe_2(As_{0.67}P_{0.33})_2$ [277]. Also the data for

[1] Note that within the sample series grown in the IFW Dresden, the sample with a nominal fluorine doping of $x = 0.04$ is not superconducting, but displays a structural and a magnetic phase transition at $T_s = 151\,K$ and $T_N = 120\,K$, respectively [135]. The $LaO_{0.96}F_{0.04}FeAs$ sample displayed in Fig. 7.1 was grown in Hosono's group [168].

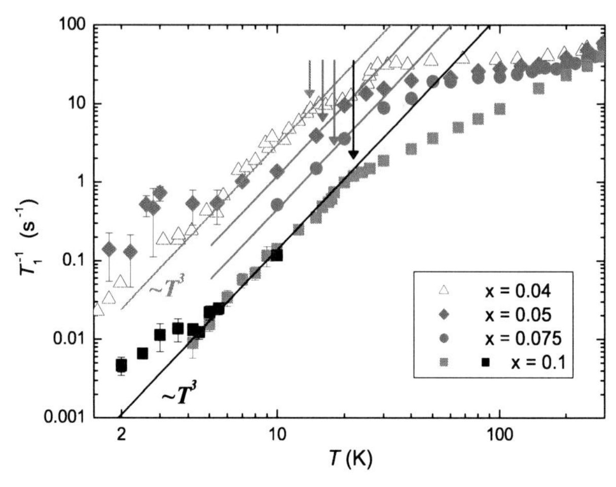


**Figure 7.1:** Temperature dependence of $^{75}$As $T_1^{-1}$ for $LaO_{0.96}F_{0.04}FeAs$ (light grey triangles, reproduced from [168]), $LaO_{0.95}F_{0.05}FeAs$ (blue rhombi), $LaO_{0.925}F_{0.075}FeAs$ (green dots) and $LaO_{0.9}F_{0.1}FeAs$ (dark grey squares, reproduced from [79] and black squares, new data). Arrows denote $T_c(H_0)$ which are approximately 14 K, 16 K, 18 K and 22 K for $x = 0.04$, 0.05, 0.075 and 0.1, respectively. Solid lines indicate $T_1^{-1} \propto T^3$. The data are measured for $H \parallel ab$. Besides the data for $LaO_{0.96}F_{0.04}FeAs$ ($H_0 = 9.9$ T), all data have been measured in $H_0 = 7.0494$ T.

$LaO_{0.96}F_{0.04}FeAs$ [168] reproduced in Fig. 7.1 show a slight deviation from the reported $T^3$ behavior at low temperatures. For $BaFe_2(As_{0.67}P_{0.33})_2$, the linear temperature behavior of $T_1^{-1}$ at low temperatures was explained with the existence of a residual density of states (RDOS) in a line-node model of the superconducting gap function [277]. This is the case of so-called "dirty *d*-wave" superconductors, where impurities lead to an additional pair breaking, smear out the line nodes and cause additional relaxation channels for the nuclear spins with a linear temperature dependence. Further evidence for line nodes in the superconducting gap function of $BaFe_2(As_{0.67}P_{0.33})_2$ came from measurements of the magnetic penetration depth [280]. However, such a high RDOS can be excluded by penetration depth data derived from $\mu$SR for the $LaO_{1-x}F_xFeAs$ samples [135] as well as for other "1111" and $Ba_{1-x}K_xFe_2As_2$ pnictide systems by several experimental measurements of the magnetic penetration depth in these systems [151–155]. Evidence for nodal superconductivity was also found in LaOFeP ($T_c \approx 6$ K), heavily overdoped $KFe_2As_2$ ($T_c \approx 3.5$ K) [169, 281, 282] and in the overdoped regime of $Ba(Fe_{1-x}Ni_x)_2As_2$ with $x \geq 0.072$ ($T_c \leq 7.5$ K) [283]. The difference in the experimental results on different pnictides were theoretically considered to stem from a switching from nodal to nodeless superconductivity upon changing the pnictogen heigth [284]. For instance, phosphorus substitution for arsenic in $BaFe_2(As_{1-x}P_x)_2$ reduces the pnictogen height, in contrast to potassium substitution for barium in $Ba_{1-x}K_xFe_2As_2$ [268]. However, only low superconducting transition temperatures are expected in the case of nodal gap functions [284].

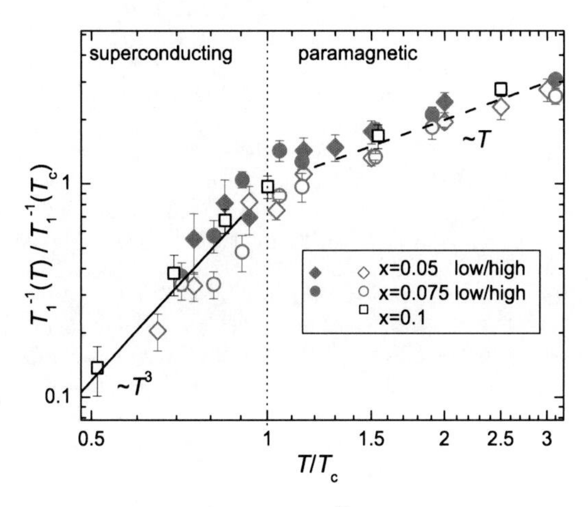


**Figure 7.2:** Temperature dependence of the $^{75}$As NQR spin-lattice relaxation rate for $LaO_{1-x}F_xFeAs$ with $x = 0.05$ (blue rhombi), $x = 0.075$ (green dots) and $x = 0.1$ (black squares). Open and closed symbols correspond to measurements on the low and high frequency resonance peak of the double-peaked NQR spectra [30]. Temperature is scaled by $T_c$ and $T_1^{-1}$ is scaled by $T_1^{-1}(T_c)$ to ease comparison. The dashed line denotes the linear temperature dependence of $T_1^{-1}$ in the paramagnetic state, while the solid line marks $T_1^{-1} \propto T^3$. The dotted line denotes $T_c$. These data are published in [30].

The relatively high $T_c$ found in $BaFe_2(As_{0.67}P_{0.33})_2$ is therefore still puzzling. Another striking difference between LaOFeP and $BaFe_2(As_{0.67}P_{0.33})_2$ on one side and LaOFeAs and $Ba_{1-x}K_xFe_2As_2$ on the other side is the dimensionality of their hole pockets. While LaOFeAs and $BaFe_2As_2$ possess two-dimensional Fermi surfaces, LaOFeP and $BaFe_2P_2$ feature a pronounced three dimensional hole Fermi pocket [268, 285]. The electron pockets remain nearly unchanged upon phosphorus substitution. The reduction of the pnictogen height upon phosphorus substitution boosts the three dimensionality of the hole Fermi surface. This weakens nesting conditions along $(\pi, \pi)$ [268]. This directly leads to another theoretical approach, which discusses the competition between orbital and spin fluctuations as a major player to explain the rich variety of the observed gap structures [286]. The strong sensitivity of the dimensionality of the hole Fermi surfaces on the doping-dependent pnictogen height might therefore play a crucial role for the existence of nodal and nodeless superconductivity in pnictides.

Measurements of the $^{75}$As NQR spin-lattice relaxation rate for $LaO_{1-x}F_xFeAs$ with $x = 0.05$, $x = 0.075$, and $x = 0.1$, are shown in Fig. 7.2. *These measurements have been performed by G. Lang and are published in [30].* Also in this case, the spin-lattice relaxation rate decreases rapidly below $T_c$ with roughly $T_1^{-1} \propto T^3$ and does not exhibit a Hebel-Slichter coherence peak. A suppression of the Hebel-Slichter peak by magnetic field, as was observed in the case of $A_3C_{60}$ [96], can thus be ruled out by the NQR measurements.

The $^{75}$As NQR spectrum of the underdoped samples with $x = 0.05$ and $x = 0.075$ consists of two broad peaks, evidencing the existence of two charge environments in the underdoped regime [30]. The $^{75}$As NQR spin-lattice relaxation rates for these underdoped samples were measured on both peaks of the double-peaked $^{75}$As NQR spectrum. The very moderate difference in absolute $T_1^{-1}$ values of the high and low frequency peak indicates that the two charge environments probed by the $^{75}$As NQR spectrum coexist at the nanoscale.

The absence of the Hebel-Slichter peak and the robust power law temperature dependence of $T_1^{-1}$ in the superconducting state stand in stark contrast to results of other experimental techniques, such as ARPES, thermal conductivity, Andreev reflection or microwave penetration depth studies, which are consistent with nodeless, fully-gapped Fermi surfaces in the superconducting state [151–159].

A reconciliation between the $T^3$ dependence of $T_1^{-1}$ in the superconducting state with measurements reporting fully gapped, nodeless Fermi surfaces is possible by considering the so-called $s_\pm$-wave symmetry (also called *extended s-wave* symmetry) for the superconducting order parameter of Fe-based superconductors, proposed by *Mazin et al.* [121]. This model accounts for the multiple Fermi surface sheets in the electronic structure of the Fe-based superconductors. The inset of Fig. 7.3 shows a sketch of a simplified Fermi surface geometry, consisting of one hole pocket at $(0, 0)$ and an electron pocket at $(\pi, \pi)$ separated by the antiferromagnetic wave vector $\vec{Q} = (\pi, \pi)$ in the folded Brillouin zone. As reported in Chapter 4, the real Fermi surface map consists of two electron pockets and, depending on the doping range, two to three hole pockets. In the undoped material, hole and electron pockets (blue and dashed blue circles) are of equal size and the good nesting conditions between them favour a magnetically ordered SDW ground state. Upon doping, the electron pocket enlarges and the hole pocket shrinks (black circles), nesting is destroyed, and superconductivity emerges. Within the proposed $s_\pm$ symmetry, the superconducting gaps which open at each Fermi surface upon entering the superconducting state are isotropic and similar in magnitude, but differ in their sign ($+\Delta$ and $-\Delta$ on the hole and the electron pocket, respectively) [121]. The pairing interaction is thus repulsive (but still leads to pairing due to the sign change). The presented spin-lattice relaxation rate data can be explained within this symmetry considerations. Based on effective two-band models like the one plotted in the inset of Fig. 7.3, early theoretical works focussed mainly on two aspects [175–177]:

**1.** As described in detail in Section 3.2.4, the sign change between the two Fermi surfaces leads to a sign reversal in the coherence factors in comparison to conventional BCS coherence factors given in Equations 3.27 and 3.28 and thus to a reduction or even disappearance of the Hebel-Slichter coherence peak already in the clean limit [103, 121, 175, 176].

**2.** To describe the observed power law dependencies of $T_1^{-1}$ below $T_c$, additionally the effect of impurities had to be taken into account. For a conventional BCS $s$-wave superconductor, non-magnetic impurities are not pair-breaking and thus do not affect the superconducting properties (*Anderson's theorem*, [287]). In the case of an $s_\pm$ symmetry this holds only true for non-magnetic impurity scattering with small scattering vectors $\vec{q}$ within a band (*intraband* impurity scattering) [121, 175, 176]. Due to the sign change of the order parameter, *interband* impurity scattering of non-magnetic impurities between electron and hole pockets with large $\vec{q}$-vectors acts like normal magnetic impurity scattering in conventional BCS $s$-wave superconductors [288]. It is pair-breaking and creates a

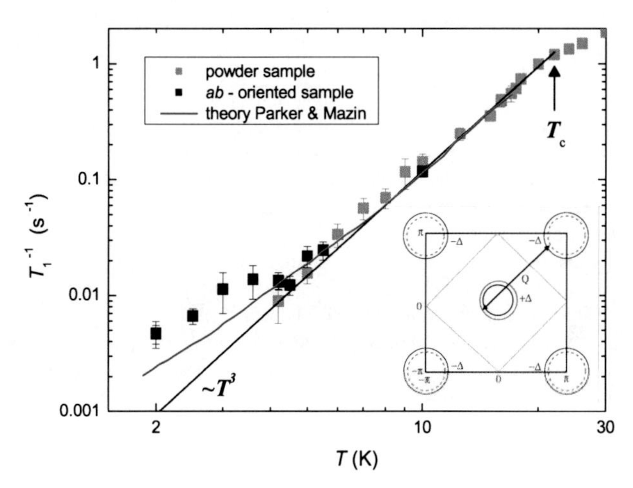


**Figure 7.3:** Temperature dependence of $^{75}$As $T_1^{-1}$ for optimally-doped $LaO_{0.9}F_{0.1}FeAs$, measured along $ab$ in $H_0 = 7.0494\,T$. Grey squares are reproduced from [79]. The arrow denotes $T_c(H_0) \approx 22\,K$. The solid black line indicates $T_1^{-1} \propto T^3$. The red solid line describes the temperature dependence of $T_1^{-1}$ in the superconducting state as theoretically expected within an $s_\pm$ symmetry of the superconducting gap for an intermediate scattering limit ($\sigma = 0.4$) and an interband impurity scattering of $\gamma_{\text{interband}} = 0.8\Delta_0$ [176]. The inset shows the simplified Fermi surface geometry based on the two-band model which was used for the calculations of $T_1^{-1}$ (reproduced from [175]).

finite density of states below the gap [121, 175–177]. Considering different ranges of impurity potentials and impurity concentrations, different power law dependencies $T_1^{-1} \propto T^n$ with $n = 2 - 3$ can be found in the superconducting state for $T < T_c$, with a further change of slope in the very low temperature region $T \ll T_c$ [175–177].

Fig. 7.3 reproduces the $^{75}$As NMR spin-lattice relaxation rate of optimally-doped $LaO_{0.9}F_{0.1}FeAs$ measured for $H \parallel ab$ in a static magnetic field of $H_0 = 7.0494\,T$, which was already depicted in Fig. 7.1. The red line is the theoretically calculated temperature dependence of $T_1^{-1}$ in the superconducting state within the previously introduced two-band $s_\pm$-wave scenario [176], considering an intermediate impurity strength[2] of $\sigma = 0.4$ and a rather strong interband impurity scattering rate[3] of $\gamma_{\text{interband}} = 0.8\Delta_0$, where $\Delta_0$ is the superconducting energy gap at low temperature without impurity scattering [176]. The data and the theoretical curve agree well in a broad temperature range below $T_c$. Even the tendency to deviate from the $T^3$ dependence at low temperature can be reproduced by theory, although absolute experimental values are higher than the theoretical curve at

[2] The impurity strength is given by $\sigma = \frac{(\pi N(0)v)^2}{1+(\pi N(0)v)^2}$ where $N(0)$ is the density of states at the Fermi energy and $v$ is the impurity potential [176].

[3] The impurity scattering rate is given by $\gamma = \frac{2c\sigma}{\pi N(0)}$, where $c$ is the impurity concentration and $\sigma$ is the impurity strength [176].

lower temperature. The value of $\gamma_{\text{interband}} = 0.8\Delta_0$ suggests a rather dirty superconductor. Note that $\gamma$ depends on the impurity concentration $c$ as well as on the impurity strength $\sigma$. Low impurity concentrations with high scattering strengths might therefore effect the scattering similarly as high impurity concentrations with low scattering strengths. In the calculations shown in Fig. 7.3, an intermediate impurity scattering of $\sigma = 0.4$ was chosen. A distribution of $\sigma$'s, which refers to the existence of different impurities with different scattering strengths, leads to an effective enhancement of $\gamma$ [176]. The high value of $\gamma = 0.8\Delta_0$ should therefore not be taken straightforwardly as an indication of high impurity concentrations.

Similar calculations which based on the five-band model proposed by Kuroki *et al.* [122] and included anisotropic $s_\pm$ gap functions, arrived at $T_1^{-1} \propto T^3$ already in the clean limit, without the necessity of strong impurity effects [289].

In conclusion, the observed peculiar behavior of the spin-lattice relaxation rate in the superconducting state can be well explained within the unconventional $s_\pm$-wave gap symmetry by considering impurity scattering effects [175–177] or by including anisotropies in the $s_\pm$ gap function [122]. These theoretical considerations were able to reconcile the NMR measurements with the observations of other experimental techniques.

## 7.2 The Effect of 'Smart' Deficiencies: $LaO_{0.9}F_{0.1}FeAs_{1-\delta}$

The NMR data reported in this Section are published in [279]. As already pointed out in the previous Section, the influence of disorder on the superconducting properties of a compound can reveal a lot about the symmetry of the superconducting state itself. By introducing disorder in a controlled way one can get an insight into relevant scattering processes. For the optimally-doped $LaO_{0.9}F_{0.1}FeAs$, such defects were produced artificially by removing some arsenic, producing As-deficiencies as a special kind of impurities. The

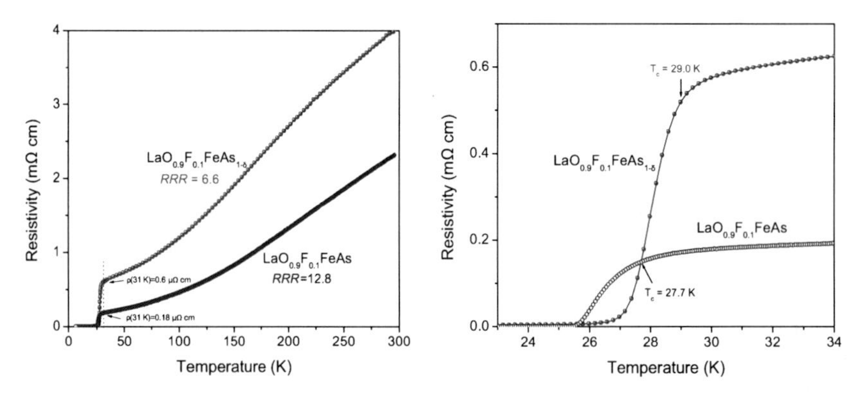


**Figure 7.4:** Resistivity of $LaO_{0.9}F_{0.1}FeAs_{1-\delta}$ (red circles) and $LaO_{0.9}F_{0.1}FeAs$ (blue circles) for temperatures up to 300 K (left panel). The right panel shows the resistivity of both samples in the vicinity of $T_c$ (reproduced from [290]).

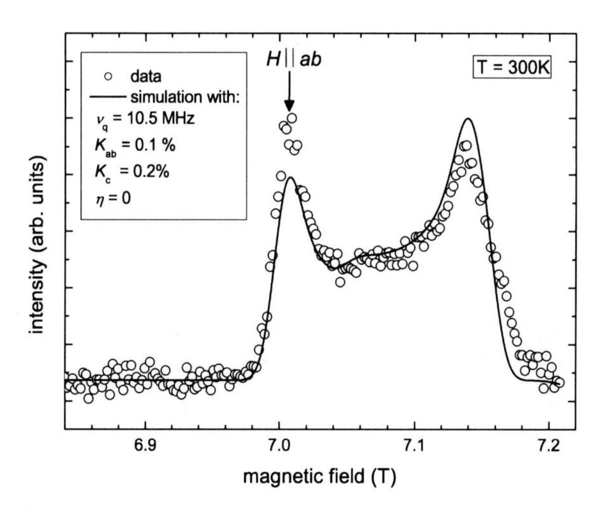


**Figure 7.5:** $^{75}$As NMR powder spectrum of $LaO_{0.9}F_{0.1}FeAs_{1-\delta}$ for the central line at room temperature. The open symbols represent the measured data, the solid line is a simulation of the powder pattern including second order electric quadrupole shifts and anisotropic magnetic hyperfine corrections. The values obtained for $\nu_q$ and $K_{ab}$ are consistent with what have been measured on $LaO_{0.9}F_{0.1}FeAs$. The slight discrepancy in intensities between the two peaks at high and low field values might stem from a partial *ab*-alignment of the crystallites.

As-deficiency was obtained by wrapping the sample in a Ta foil during the annealing procedure. At high temperature Ta acts as an As getter. It forms a solid solution of about 9.5 at% As in Ta with a small layer of $Ta_2As$ and TaAs on top of the Ta foil. According to energy dispersive x-ray (EDX) analysis, this leads to an As-deficiency in the wrapped sample of about $\delta$=0.05-0.1. An analysis of the NQR spectrum found the amount of arsenic vacancies to be $\delta = 0.06 \pm 0.02$, in agreement with EDX studies [291]. The resulting sample was further characterized by x-ray diffraction, susceptibility and resistivity measurements [219, 290].

The increased disorder in $LaO_{0.9}F_{0.1}FeAs_{1-\delta}$ is reflected in an enhanced resistivity in the normal state compared to a clean sample of $LaO_{0.9}F_{0.1}FeAs$ [179] (see Fig. 7.4). Despite the increased disorder, $T_c$ and the slope of $B_{c2}(T)$ near $T_c$ *increase* unexpectedly from 27.7 K and -2.5 T/K in the clean compound to 28.5 K and -5.4 T/K in the As-deficient compound. The growth procedure, where vacancies start to be formed at the surface of the sample, could suggest an inhomogeneous distribution of arsenic vacancies in the sample. However, the sharp superconducting transition (see inset of Fig. 7.4), the shape of the NQR spectrum [291] and the surprisingly strong decrease of $T_1^{-1}$ with $T^5$ in the superconducting state, which will be reported below, indicate a homogeneous distribution of arsenic vacancies in the compound. This might be related to the strongly anisotropic structure of LaOFeAs, which prevents the clustering of vacancies [291].

$\mu$SR measurements proved an enhanced paramagnetism, which is the origin of the observed Pauli-limiting behavior of $B_{c2}(T)$ at lower temperatures [290]. The supercon-

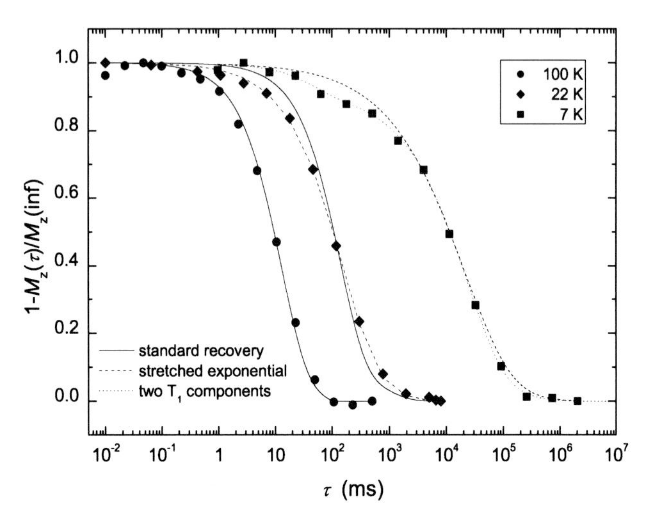


**Figure 7.6:** Recovery curves for $T = 100$ K (dots), $T = 22$ K (diamonds) and $T = 7$ K (squares) in $H_0 = 7.01$ T. Normalization corresponds to a division by the prefactor $f$(=1.7-2) of Eq. (7.1). The lines are examples for the different fitting functions containing a single $T_1$ component (solid line), a distribution around one $T_1$ component with a stretching parameter $\lambda$ (dashed line) and two components $T_{1sc}$ and $T_{1s}$ (dotted line).

ducting volume fraction is about 90% while in the pure sample it amounts to 100% [135]. The enhanced paramagnetism was confirmed by observations of a significantly enhanced spin susceptibility in the As-deficient sample in comparison to clean $LaO_{0.9}F_{0.1}FeAs$ [291].

$^{75}$As NMR measurements were performed on a powder sample of $LaO_{0.9}F_{0.1}FeAs_{1-\delta}$. The $^{75}$As NMR spectrum, displayed in Fig. 7.5, shows a typical powder pattern as reported previously [79] (see App. A.1 for a detailed discussion of NMR powder pattern). The $^{75}$As spin-lattice relaxation rate $T_1^{-1}$ was measured at the peak corresponding to $H \| ab$ in a magnetic field of $H_0 = 7.01$ T using inversion recovery. The recovery of the longitudinal magnetization $M_z(t)$ was fit to the standard expression for magnetic relaxation of a nuclear spin of $I = 3/2$ which reads:[4]

$$M_z(t) = M_0[1 - f(0.9\mathrm{e}^{-(6\mathrm{t}/\mathrm{T}_1)^\lambda} + 0.1\mathrm{e}^{-(\mathrm{t}/\mathrm{T}_1)^\lambda})]\,. \tag{7.1}$$

Typical recovery curves for $T = 100$ K, $T = 22$ K and $T = 7$ K are shown in Fig. 7.6. Above $T_c(H_0) \approx 26.5$ K the recovery could be nicely fit with a single $T_1$ component ($\lambda = 1$). For $T < T_c$ a stretching parameter $\lambda < 1$ was needed to account for a distribution of spin-lattice relaxation times around a characteristic relaxation time. For $T \leq 14$ K, where the intrinsic spin-lattice relaxation time $T_{1sc}$ in the SC state amounts to a few

[4] Please refer to Appendix A.3 for a discussion of the use of the stretching exponent $\lambda$.

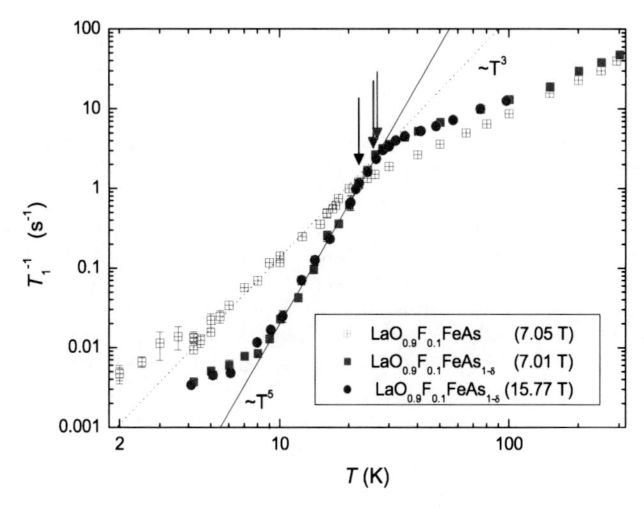


**Figure 7.7:** $^{75}$As spin-lattice relaxation rate for $LaO_{0.9}F_{0.1}FeAs_{1-\delta}$ in 7.01 T (red squares) and 15.77 T (blue circles) compared to $LaO_{0.9}F_{0.1}FeAs$ in 7.05 T (grey crossed squares [79], new data points for $T \leq 4.2$ K). All data were measured for $H \parallel ab$. The dotted line illustrates the $T^3$ behavior of $T_1^{-1}$ for $LaO_{0.9}F_{0.1}FeAs$, the solid line indicates the $T^5$ behavior observed for $LaO_{0.9}F_{0.1}FeAs_{1-\delta}$. The arrows denote $T_c(H_0)$ which are approximately 22 K for $LaO_{0.9}F_{0.1}FeAs$, and 26.5 K (7.01 T) and 25 K (15.77 T) for $LaO_{0.9}F_{0.1}FeAs_{1-\delta}$.

seconds, one could distinguish a second, short contribution $T_{1s}$. For this temperature-range a fitting function containing two weighted $T_1$ components was used:

$$\begin{aligned} M_z(t) &= M_0[w(1 - f_{sc}(0.9\mathrm{e}^{-(6t/T_{1sc})^{\lambda_{sc}}} + 0.1\mathrm{e}^{-(t/T_{1sc})^{\lambda_{sc}}})) + \\ &\quad (1-w)(1 - f_s(0.9\mathrm{e}^{-(6t/T_{1s})^{\lambda_s}} + 0.1\mathrm{e}^{-(t/T_{1s})^{\lambda_s}}))] . \end{aligned} \tag{7.2}$$

During the fitting procedures, $\lambda_s$ was always kept constant at $\lambda_s = 1$, since the short component $T_{1s}$ stems from non-superconducting regions, where a distribution of spin-lattice relaxation rates is not expected. $\lambda_{sc}$ varied between 1 and 0.6. The exact determination of $T_{1s}$ was imprecise. More importantly, the long time component $T_{1sc}$, which displays the intrinsic relaxation in the superconducting state, did not depend on the fitting procedure.[5] $T_{1s}$ lies in the range of several hundred ms, indicating non-SC regions in the sample. Its weight $w_s = (1-w) \approx (20 \pm 10)\,\%$ suggests, in addition to vortex cores, a minor non-superconducting volume fraction in the sample, in agreement with $\mu$SR-measurements [290].

[5] Since Eq. (7.2) contains 8 parameters, the robustness of the fit was checked by fixing different parameters to reasonable values during different fitting sessions. For instance at a first run $w$ was fixed to 0.81 and at a second run $T_{1s}$ was fixed to 220ms. $T_{1sc}$ resulted to be the same within error bars for all different checks. During the first run, $T_{1s}$ lay in between 100 and 450 ms and during the second check $w$ varied between 0.65 and 0.81, which are realistic values.

Fig. 7.7 shows the $T$-dependence of the $^{75}$As spin-lattice relaxation rate $T_1^{-1}$ in $H_0 = 7.01\,T$ for the As-deficient sample $LaO_{0.9}F_{0.1}FeAs_{1-\delta}$ [279] and that of a sample with the same fluorine content, but without As-vacancies [79] (same data as in Fig. 7.1 and Fig. 7.3). Very surprisingly, the spin-lattice relaxation rate of $LaO_{0.9}F_{0.1}FeAs_{1-\delta}$ decays with $T^5$ for $T < T_c$, in stark contrast to the $T^3$ dependence of $LaO_{0.9}F_{0.1}FeAs$. Using a field of 15.77 T this unexpected behavior was preserved within error bars, as also shown in Fig. 7.7.

For $T \leq 8$ K, $T_1^{-1}$ of $LaO_{0.9}F_{0.1}FeAs_{1-\delta}$ deviates from this $T^5$ behavior and changes to a linear temperature dependence. As already discussed in the previous Section, such a nearly linear slope below $T \approx 0.3T_c$ was also observed in $BaFe_2(As_{0.67}P_{0.33})_2$ and explained with the existence of a residual density of states (RDOS) in a line-node model [277], which can be excluded by penetration depth data derived from $\mu$SR for the La1111 samples [135]. Among other possible mechanisms, the classical spin diffusion [292] is unlikely due to the lack of field dependence of $T_1^{-1}$ (see also Appendix A.4). Another possibility are thermal fluctuations of vortices, which induce alternating magnetic fields contributing to the relaxation [293]. The effect of thermal fluctuations on the spin-lattice relaxation is field-independent as long as $H \ll H_{c2}$.

The different $T$-dependencies for $T > 0.3T_c$ will now be discussed within the previously mentioned $s_\pm$ pairing scenario.

Up to now, with the exception of $LaO_{0.9}F_{0.1}NiAs$, which exhibits an exponential decrease (see Chapter 4), no exponential but power-law dependencies $T_1^{-1} \propto T^n$ have been observed for all Fe-based superconductors, with $n$ in between 1.5 and 6, indicating unconventional superconductivity [30, 79, 147–150, 162, 164–173, 277, 279, 294]. These power law dependencies have been discussed within different models, such as $s_\pm$- and $d$-wave symmetries (see also previous Section), including the possibility of multiple superconducting gaps [147, 165, 170, 171]. Within the 122 family heavily overdoped compounds such as $KFe_2As_2$ exhibit the lowest value observed so far whereas optimally or slightly underdoped compounds show the largest values. Recently, it has been suggested [162, 173] that the most frequently observed $T^3$ dependence of $T_1^{-1}$ should not be considered as an intrinsic effect but instead be attributed to some unspecified inhomogeneities in view of the missing correlation between the value of $T_c$ and the exponent, while higher exponents $n$ would occur for cleaner samples. However, in the case of $LaO_{0.9}F_{0.1}FeAs_{1-\delta}$ just the opposite behavior is observed. Fig. 7.8 shows the normalized $T_1^{-1}(T)/T_1^{-1}(T_c)$ curves for the As-deficient sample and the samples from Ref. [173]. Their nominal clean sample as well as the Co-doped one exhibit nearly the same $T_1^{-1}(T)$-dependence as the As-deficient sample, whereas our clean sample exhibits the $T^3$ dependence (see Fig 7.7). This points towards sizeable disorder in the samples of Ref. [173]. This is further supported by the lowest resistivity of the clean sample from the IFW compared to all others [173, 290].

In principle, the observation of an unusual transition from $T^3$ to $T^5$ with *increasing* disorder is not necessarily inconsistent with a $s_\pm$-wave superconducting gap function though alternative scenarios should be invoked, too. Starting from the clean limit it has been shown [115, 284, 295] that within the generalized $s_\pm$-wave scenario both nodeless and nodal superconducting gaps might occur depending on the proximity of the doped sample to the antiferromagnetic instability. In this regard, naively the results can be interpreted in favour of a transition from a nodal to a nodeless superconducting gap upon adding As defects which for some reason might drive the system closer to antiferromagnetism,

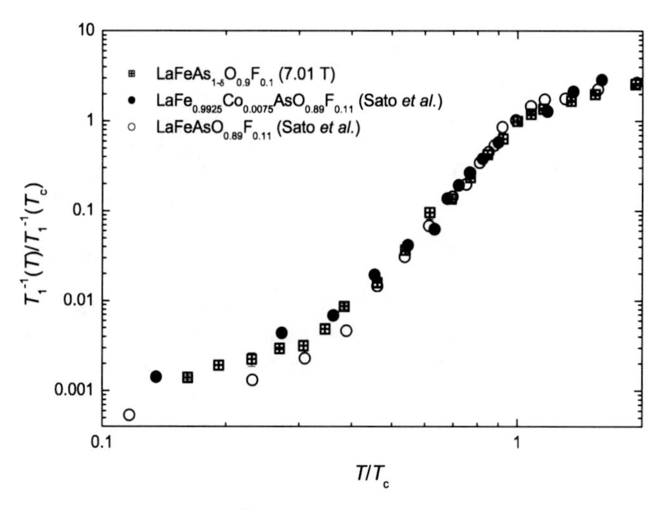


**Figure 7.8:** Comparison of the $^{75}$As spin-lattice relaxation rate for our $LaO_{0.9}F_{0.1}FeAs_{1-\delta}$ (crossed squares) with pure and Co-doped $LaFeAsO_{0.89}F_{0.11}$ from Ref. [173] (open and filled circles).

in accord with the slightly enhanced normal state spin-lattice relaxation rate and the slightly changed lattice constants [290] of the As deficient sample. However, such a simplistic point of view cannot be easily applied to pnictides as it is also known that the $s_\pm$-wave ground state is sensitive to non-magnetic impurities (see discussion in the previous Section). Most importantly, the *intraband* impurity scattering does not affect the superconductivity, since the superconducting gap does not change its sign within each of the bands. At the same time the scattering with large momenta which connects electron and hole pockets (*interband* scattering) is pair-weakening and thus yields a decrease of $T_c$ and simultaneously introduces power laws in the thermodynamics and $T_1^{-1}$ at intermediate temperatures. Therefore, if for some reason As vacancies act as 'smart' impurities which change the ratio between the *intra-* and *interband* scattering, our observations could be also explained. These changes have to be reflected similarly also in the other thermodynamical or transport properties such as penetration depth or thermal conductivity. Unfortunately, there is no direct way to estimate the ratio of the *intraband* to *interband* scattering rates from experiments since usual characteristics like the residual resistance ratio or the mean free path are quantities which mostly indicate the overall impurity effects but not their ratio.

Thus, the above-mentioned scenario is based on the assumption that $s_\pm$-wave order is stable and adding As vacancies either changes the proximity to the competing antiferromagnetism or/and the ratio of *intra-* to *interband* non-magnetic impurity scattering in pnictides. There is, however, another intriguing possibility. Let us assume that there is a substantial electron-boson interaction which provides an attractive *intraband* potential for Cooper-pairing. In this case a (weak) repulsive *interband* Coulomb scattering will still lead to the $s_\pm$-wave SC order in the clean limit though the attractive electron-boson interaction dominates. However, once the As vacancies change the ratio between *intra-* and *interband* impurity scattering, a transition from $s_\pm$-wave to conventional $s_{++}$-wave superconducting order may be induced. This scenario, however, still needs further experimental clarification. For example, despite the transition from $T^3$ to $T^5$ behavior no sign of the Hebel-Slichter peak is found in the latter case close to $T_c$. Moreover, current experimental data on the importance of the electron-phonon coupling are not very conclusive. Therefore, the intriguing possibility of high-energy charge fluctuations as well as weak electron-phonon interactions with orbital fluctuations [286, 296] deserve more detailed studies.

All previously discussed scenarios based on the assumption, that arsenic vacancies act as non-magnetic impurities in the sample, being pair-breaking for *interband* scattering processes with large momenta $\vec{q}$. However, recently it has been shown that the enhanced spin susceptibility of $LaO_{0.9}F_{0.1}FeAs_{1-\delta}$ can be attributed to the formation of non-compensated magnetic moments around each arsenic vacancies of about $3.2\,\mu_B$ per vacancy ($0.8\mu_B$/Fe) [291]. The higher spin-lattice relaxation rate of $LaO_{0.9}F_{0.1}FeAs_{1-\delta}$ in the normal state compared to $LaO_{0.9}F_{0.1}FeAs$ (see Fig. 7.7) is consistent with this local moment picture. It might by enhanced by a contribution related to these local moments which form around each arsenic vacancy. On the other hand, magnetic impurities are harmless for a superconducting state exhibiting an $s_\pm$ symmetry, but pair-breaking for a more conventional $s_{++}$ superconducting symmetry. In this regard, the faster decrease of $T_1^{-1}$ in the superconducting state of $LaO_{0.9}F_{0.1}FeAs_{1-\delta}$ might be taken as a strong evidence for a stable $s_\pm$ symmetry. In this case, the absence of the Hebel-Slichter peak is a natural consequence of the gap symmetry itself (as discussed in the previous Section).

## 7.3 Summary

$^{75}$As NMR and NQR measurements in the superconducting state of underdoped and optimally-doped $LaO_{1-x}F_xFeAs$ with $x = 0.05$, $x = 0.075$ and $x = 0.1$ showed a rapid decrease of the spin-lattice relaxation rate proportional to $T^3$ without any hint for a Hebel-Slichter coherence peak. At low temperatures a deviation to a more linear temperature dependence was observed. The data are consistent with NMR measurements by other groups on the same and similar compounds. At the beginning of the "pnictide - era", the robust power law dependence of $T_1^{-1}$ was taken as an indication of the existence of line nodes in the superconducting gap function. This stood in contrast to the observation of fully gapped, nearly isotropic gap functions by the means of ARPES, microwave penetration depth studies and Andreev reflection.

A revision of theoretical approaches was given. These are able to reconcile the results of different experimental methods within the $s_\pm$ wave symmetry of the superconducting gap function, where the (mostly) isotropic gap changes its sign when going from an electron to a hole pocket and vice versa. Due to the sign changing order parameter, the Hebel-Slichter peak is reduced or even totally suppressed. Power law dependencies of $T_1^{-1}$ can be achieved theoretically by considering the pair-breaking effect of *interband* impurity scattering effects of non-magnetic impurities with large $\vec{q}$-vectors within a simplified two-band model [175–177] or by the consideration of anisotropic gap functions within a five-band model approach in the clean limit [289].

The effect of impurities was studied on an optimally-doped sample with additional arsenic vacancies, $LaO_{0.9}F_{0.1}FeAs_{1-\delta}$. Surprisingly, despite the enhanced disorder, this compound possesses enhanced superconducting properties (higher $T_c$, higher $H_{c2}$, ...) [219, 290]. Its spin-lattice relaxation rate decreases proportional to $T^5$ and thus much faster than the one of the clean optimally-doped $LaO_{0.9}F_{0.1}FeAs$ sample. This enhanced decrease was discussed within three scenarios, which all based on the assumption, that the arsenic vacancies act as non-magnetic impurities: an impurity-induced crossover from nodal to nodeless superconductivity within the $s_{\pm}$-wave symmetry state, a stabilization of the $s_{\pm}$ symmetry due to an impurity-induced change of the ratio between *intra-* and *interband* scattering, and an impurity-induced crossover from an $s_{\pm}$ to an $s_{++}$ symmetry. However, as has been shown in a recent publication [291], local moments form around each arsenic vacancy. A reinterpretation of the $T_1^{-1}$ data, considering arsenic vacancies as effective magnetic impurities gives evidence for an unconventional $s_{\pm}$ symmetry of the superconducting order parameter.

# 8 NMR and NQR on LiFeAs

This Chapter will present $^{75}$As NQR, $^{75}$As NMR and $^{7}$Li NMR measurements on three different single crystals and a polycrystalline sample of LiFeAs. As will be shown, no common behavior could be found. Static and dynamic NMR properties vary among the samples, which are distinguishable by their slightly different quadrupole frequencies. Samples with a lower quadrupole frequency $\nu_q$ exhibit a decrease of the Knight shift and the spin-lattice relaxation rate in the superconducting state, indicating usual spin-singlet superconductivity. For the sample with the highest $\nu_q$, a constant Knight shift and a strange upturn of the $^{75}$As NQR spin-lattice relaxation rate in the superconducting state provide evidence for spin-triplet superconductivity. Surprisingly, all samples seem to be very homogeneous, according to their very narrow $^{75}$As NQR and $^{7}$Li NMR spectra and other physical properties (please refer to Section 5.3.2 for a detailed discussion of the sample quality). These results provide evidence for the proximity of LiFeAs to a ferromagnetic instability, where tiny changes in the stoichiometry, seen only by NQR, can lead to different Cooper pairing states. As will be discussed, these data are in agreement with recent band structure calculations. By including special characteristics of LiFeAs measured by ARPES, such as the absence of Fermi surface nesting and the shallow hole pockets around the $\Gamma$ point (see Fig. 4.9), these calculations found dominant "almost ferromagnetic" incommensurate fluctuations in LiFeAs and thus a proximity to a ferromagnetic instability [214].

The discussion of the data starts with the $^{75}$As NQR spectra and spin-lattice relaxation rates. Afterwards, $^{75}$As NMR linewidths, Knight shift and spin-lattice relaxation rate data will be shown and interpreted. Several possible origins of the observed constant Knight shift for sample S1 will be carefully discussed, including the possibility of spin-triplet superconductivity. The peculiar behavior of the spin-lattice relaxation rate in S1 and in the polycrystalline sample will be compared to Ca-doped LaOFeP. This compound showed a similar behavior, which is possibly related to spin-triplet superconductivity [297]. The Chapter ends with $^{7}$Li NMR results.

The single crystalline samples are denominated as S1, S2 and S3. The data of S3 have been measured by the author, *while S1 and S2 have been measured by Seung-Ho Baek and Hans-Joachim Grafe, respectively. The $^{75}$As NQR measurements on the polycrystalline sample were done by Madeleine Fuchs*, while the $^{75}$As NMR measurements on the same powder sample were done by the author. Most of the data presented in this Chapter are published in [298].

## 8.1 $^{75}$As-NQR

Fig. 8.1 shows the $^{75}$As NQR spectra of all three investigated single crystals at room temperature. Very narrow lines with full widths at half maximum (FWHM) of 44 kHz (S3),

60 kHz (S2) and 80 kHz (S1) are observed. These are significantly narrower than the $^{75}$As NQR linewidths of the polycrystalline sample (113 kHz) and of a polycrystalline sample measured by another group (170 kHz) [81]. Other undoped pnictides feature linewidths in the range of some hundred kHz, such as 220 kHz for stoichiometric LaOFeAs [30, 193] and 480 kHz for $CaFe_2As_2$ [224]. The extremely narrow NQR lines observed for our LiFeAs single crystals are even more surprising regarding the high quadrupole frequency of about 21.5 MHz compared to LaOFeAs ($\nu_q(160\,\mathrm{K}) \approx 9.5\,\mathrm{MHz}$) [30, 193] and $CaFe_2As_2$ ($\nu_q(250\,\mathrm{K}) \approx 12\,\mathrm{MHz}$) [224, 251], because a distribution of $\nu_q$ values induced by disorder should lead to an even larger linewidth. Since the effect of disorder strongly influences the quadrupole broadening (see Section 2.4), the very narrow $^{75}$As NQR linewidths reflect the high purity of all three investigated single crystals. The temperature dependence of the $^{75}$As NQR linewidth will be discussed in Section 8.2.1 in direct comparison to the $^{75}$As NMR linewidth (see Fig. 8.4).

Fig. 8.2 shows the temperature dependence of $^{75}$As NQR quadrupole frequency $\nu_q$ of S1, S2, S3 and of the polycrystalline sample. While the overall temperature dependence of $\nu_q$ in the paramagnetic phase is the same, displaying a decrease of $\nu_q$ with decreasing temperature, the absolute values of $\nu_q$ and thus the EFG varies with the sample investigated. The samples can be clearly distinguished by their different $\nu_q$ at a given temperature, pointing towards slightly different, but homogeneous charge environments for each sample. This resembles the doping dependence of $\nu_q$ in other pnictides [30, 148, 299]. Although all the LiFeAs samples are from the same batch and should have the same composi-

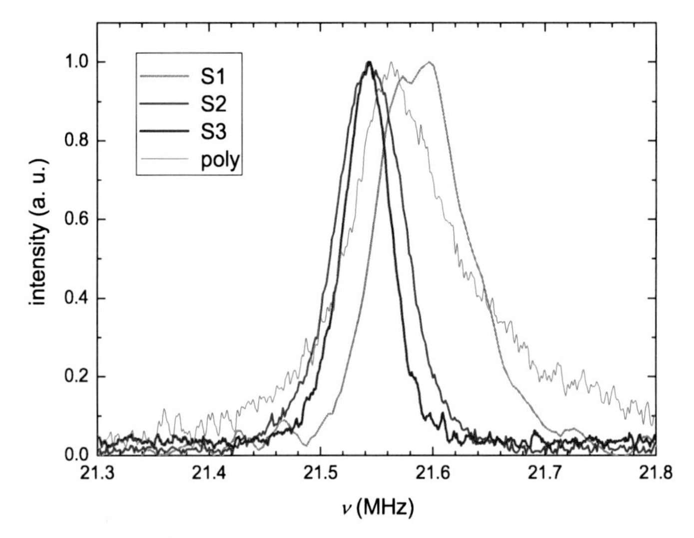


**Figure 8.1:** Normalized $^{75}$As NQR spectra of LiFeAs single crystals (S1 - grey line, S2 - red line, S3 - blue line) and the polycrystalline sample (green line) at room temperature.

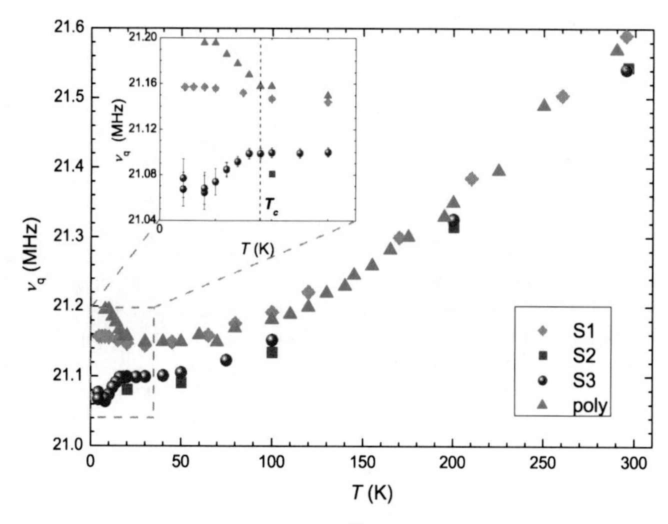


**Figure 8.2:** Temperature dependence of the $^{75}$As NQR quadrupole frequency for S1 (grey rhombi), S2 (red squares), S3 (blue dots) and the polycrystalline sample (green triangles). The inset is an enlargement of the low temperature part. The dashed line in the inset denotes $T_c = 18\,\text{K}$.

tion according to ARPES and inductively coupled plasma mass spectroscopy (ICPMS) measurements, they might slightly differ in their stoichiometries, leading to the observed different quadrupole frequencies. Interestingly, the samples with the higher $\nu_q$ (S1 and the polycrystalline sample) exhibit strange NMR properties in the superconducting state (see next Section). The quadrupole frequencies of S2 and S3 are similar and lower than the one of S1.

A strong temperature dependence of $\nu_q$, far beyond the slight enhancement of $\nu_q$ which would be expected from phononic contributions, (and mostly in the opposite direction), is a common feature in iron-based superconductors. $\nu_q$ either decreases (e.g. for $BaFe_2As_2$ [228] and $LaO_{0.9}F_{0.1}FeAs$ [300]) or increases (e.g. $CaFe_2As_2$ [251] and $SrFe_2As_2$ [301]) strongly with temperature. This stark temperature dependence is attributed to electronic effects due to the complicated multi-band character and may reflect the extreme sensitivity of the electronic structure to the out-of-plane atoms.

The inset of Fig. 8.2 shows the low temperature behavior of $\nu_q$. After a levelling off between 50 K and $T_c$ for all samples, $\nu_q$ increases in the case of S1 and the polycrystalline sample, while it decreases in S3. As will be shown in Section 8.2 a precise knowledge of the quadrupole frequency is needed to accurately correct the $^{75}$As NMR resonance frequency of the central transition for second order quadrupole effects, before extracting the Knight shift. In S2 the diamagnetic shielding of the superconducting state was so good that it was not possible to perform $^{75}$As NQR measurements in the superconducting state in this specific sample.

### $^{75}$As NQR Spin-Lattice Relaxation Rate

The $^{75}$As NQR spin-lattice relaxation rate divided by temperature, $(T_1T)^{-1}$, is displayed in Fig. 8.3 for samples S1, S3 and the polycrystalline sample. In the paramagnetic state, their absolute values are similar and they all exhibit the same temperature dependence. $(T_1T)^{-1}$ slightly decreases with decreasing temperature in the high temperature regime and then levels off at a constant value below $T \approx 150\,\mathrm{K}$. In the superconducting state, however, very different temperature dependences are observed. While $(T_1T)^{-1}$ of sample S3 drops below $T_c$, indicating the opening of the superconducting gap and the associated reduction of the electron density of states at the Fermi level, $(T_1T)^{-1}$ of sample S1 and of the polycrystal increases below $T_c$. At first glance one might suspect, that these two samples are not superconducting. However, superconductivity has been confirmed in all samples by observing the change in the resonance frequency of the NMR circuit, which is proportional to the ac susceptibility. The fact that the upturn of $(T_1T)^{-1}$ in sample S1 and in the polycrystalline sample (which are the two samples with the higher $\nu_q$) starts just below $T_c$, indicates that this upturn might be connected to the superconducting state. In Section 8.2 this enhancement of the $^{75}$As NQR $(T_1T)^{-1}$ in sample S1 and the polycrystal will be compared to the equally unusual $^{75}$As NMR $(T_1T)^{-1}$ of the same samples (see Fig. 8.8) and to a similar behavior observed in Ca-doped LaOFeP [297].

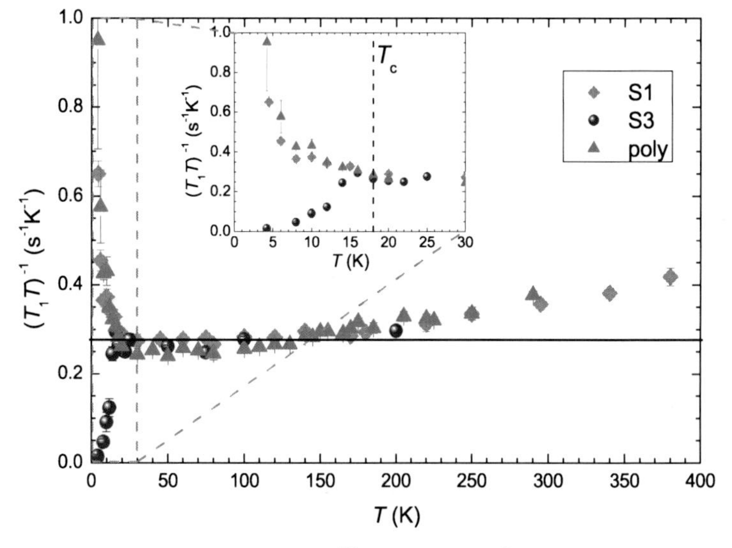


**Figure 8.3:** Temperature dependence of $^{75}$As NQR $(T_1T)^{-1}$ for sample S1 (grey rhombi), sample S3 (blue dots) and the polycrystalline sample (green triangles). The solid line is a guide to the eyes. The inset shows an enlargement of the low temperature part. The dashed line in the inset denotes $T_c = 18\,\mathrm{K}$.

# 8.2 $^{75}$As-NMR

## 8.2.1 Static NMR Properties - Linewidth and Knight Shift

$^{75}$As NMR measurements of the central transition $(m = -\frac{1}{2}) \longleftrightarrow (m = +\frac{1}{2})$ were performed for $H_0 \parallel ab$ for all three samples. For samples S1 and S3, the measurements were done in a field of $H_0 = 7\,\mathrm{T}$. For S2, the field value was $H_0 = 8.5\,\mathrm{T}$. Additionally to the measurements along $ab$, the angle dependence of the linewidth and the Knight shift was checked for samples S1 and S2, by slightly tilting the field out of the $ab$ plane by an angle $\theta$. Sample S1 was also measured for $H_0 \parallel c$.

### Linewidth

The $^{75}$As NMR spectra at room temperature are very narrow, exhibiting a full width at half maximum (FWHM) of 23.5 kHz (S3) and 28 kHz (S1) [see Fig. 8.4(b)].[1] The temperature dependence of the $^{75}$As NQR and NMR linewidths is given in Fig. 8.4(a) and 8.4(b), respectively. For S1 and S3, the $^{75}$As NQR as well as the $^{75}$As NMR linewidth (regardless of the angle) increase with decreasing temperature in the paramagnetic phase. According to Section 2.4 this implies that magnetic correlations progressively gain in strength upon lowering the temperature. At high temperature, local moments are fluctuating with a much higher frequency $1/\tau_c$ than the NMR frequency. The nuclei can only sense the time-averaged local field and the NMR line will be narrow (*motional narrowing*). As the fluctuating moments slow down, the nuclei will begin to feel a distribution of local fields. This will broaden the NMR and NQR linewidths, as observed in the case of samples S1 and S3. For sample S2, however, the $^{75}$As NQR linewidth is temperature-independent, as indicated by the red line in Fig. 8.4(a). This result together with the smallest absolute value of the $^{75}$As NMR linewidth of S2 at low temperature in comparison to all other samples also points to a temperature-independent NMR linewidth for this sample [indicated by the red line in Fig. 8.4(b)]. Thus, regarding the linewidth behavior, no hint for magnetic correlations is found in sample S2, in contrast to the other two samples.

For sample S1, the linewidth has also been measured as a function of the tilting angle $\theta$. Already a small tilting angle of 2° causes a noticeable broadening, which gets even more pronounced when increasing the tilting angle to 5.6° [see Fig. 8.4(b)]. These strong angle dependencies indicate the existence of anisotropic spin fluctuations. This is consistent with the incommensurate, "almost ferromagnetic" correlations which have been reported in recent band structure calculations [214].

In the superconducting state, the temperature dependence of the linewidth also varies with the investigated sample. For S1 it decreases below $T_c$ for both, NQR and NMR measurements. For S2 the $^{75}$As NMR linewidth increases. This is the usually expected behavior due to vortex-related broadening mechanisms in the superconducting state. The decrease of the linewidth in the case of S1 excludes possible vortex-related broadening mechanisms and suggests a suppression of spin fluctuations in the superconducting state

[1] Note that sample S2 could not be measured by $^{75}$As NMR at high temperatures. To perform such measurements, the resonant circuit had to be changed, which implied that the sample probe had to be taken out of the cryostat. During this process, the sample degraded because of its extreme sensitivity to air and moisture.

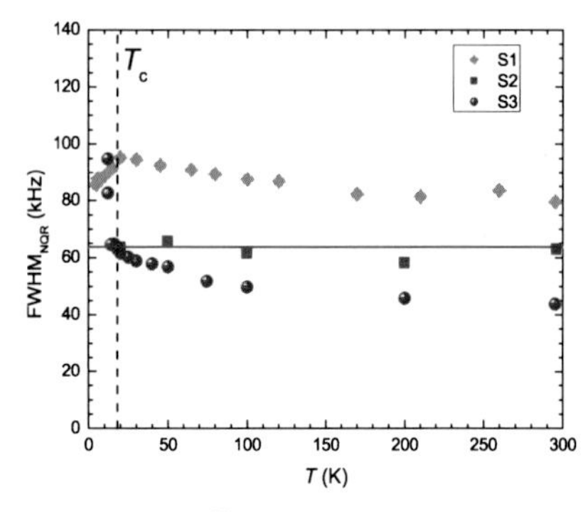


(a) FWHM of $^{75}$As NQR spectra for S1 (grey rhombi), S2 (red squares) and S3 (blue dots). The dashed line denotes $T_c = 18$ K. The red line is a guide to the eyes.

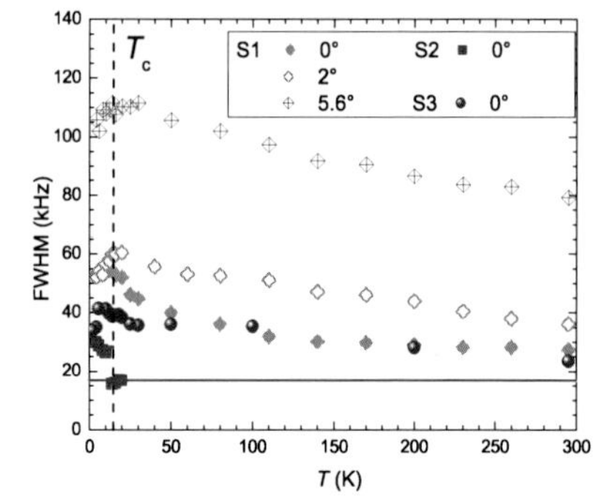


(b) FWHM of $^{75}$As NMR spectra for $H_0 \parallel ab$ for S1 (grey filled rhombi), S3 (blue dots) (both measured in 7 T) and S2 (red squares, 8.5 T). For S1, also the FWHM for small tilting angles $\theta$ of $H_0$ vs $ab$ are shown (grey open rhombi). The dashed line denotes $T_c(7\,\mathrm{T}) = 14.6$ K. The red line is a guide to the eyes.

**Figure 8.4:** FWHM of $^{75}$As NQR spectra (left panel) and $^{75}$As NMR spectra for $H_0 \parallel ab$ (right panel) for all three measured single crystals. For sample S3, additional angle-dependent FWHM are shown. To facilitate comparison, both data sets have been plotted on the same scale. The very low temperature part of the $^{75}$As NQR linewidth of S3, which goes up to 240 kHz at 4.2 K, is thus out of the scale.

(in field as well as in zero field). In the case of sample S3, the NQR linewidth increases very strongly in the superconducting state, up to 240 kHz at 4.2 K. The NMR linewidth of S3 first increases slightly, but then drops again. The origin of this very peculiar behavior is unclear. It suggests that spin fluctuations, which are present in the superconducting state in zero field, are suppressed by an externally appplied field.

### Knight Shift

To extract the Knight shift, the resonance frequency of the $^{75}$As NMR spectra first had to be corrected for second-order quadrupole effects according to Eq. (2.39). Due to the large value of $\nu_q$, the second-order quadrupole shift amounted up to $\Delta^{(2)}\nu(\theta) \approx 1.5$ MHz. The strong dependence of $\Delta^{(2)}\nu(\theta)$ on the angle $\theta$, which is the angle between the principal axis of the EFG ( which in the case of LiFeAs is $V_{zz} \parallel c$, see Section 8.4) and the direction of the magnetic field $H_0$, could be used to align the samples with great accuracy.

The resulting Knight shift data at low temperature are shown in Fig. 8.5. The left panel of Fig. 8.5, displays the angle-dependent Knight shift of samples S1 and S2. For sample S1 and $H_0 \parallel ab$ ($\theta = 0°$), the Knight shift $K_{ab}$ does not show any change upon crossing $T_c$. This is in stark contrast to what is expected and observed in spin-singlet superconductors. Even more surprising, this behavior changes when the field is slightly tilted out of the $ab$ plane of the sample. Already for a very small tilting angle of $\theta = 2°$, the

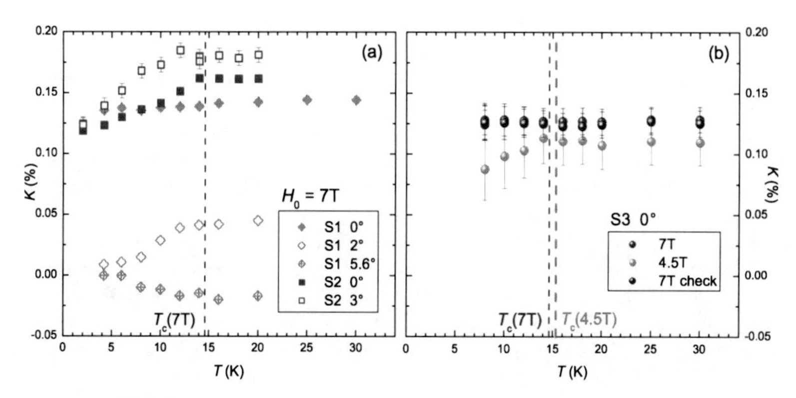


**Figure 8.5:** $^{75}$As NMR Knight shift of LiFeAs at low temperature. The left panel shows the Knight shift and its angular dependence for sample S1 in $H_0 = 7\,\mathrm{T}$ (grey rhombi) and sample S2 in $H_0 = 8.5\,\mathrm{T}$ (red squares). Filled symbols denote $H_0 \parallel ab$, open symbols denote a tilting of the field out of the $ab$ plane by the corresponding angle. The right panel shows the Knight shift for sample S3 with $H_0 \parallel ab$ for two measurements in $H_0 = 7\,\mathrm{T}$ (blue and dark red dots) and an interjacent measurement in $H_0 = 4.5\,\mathrm{T}$ (orange dots). Dashed lines denote $T_c(H_0)$ (14.6 K for 7 T and 15.3 K for 4.5 T). The same scale is chosen for both figures to facilitate comparison.

Knight shift clearly drops below $T_c$ and approaches a finite value as $T \to 0$. For $\theta = 5.6°$, the Knight shift even increases slightly below $T_c$, but approaches the same finite value as in the case of $\theta = 2°$ at low temperature. Measurements on polycrystalline samples of LiFeAs performed by other groups reported very similar data to the $\theta = 2°$ data shown here, but assigned it to a $H_0 \parallel ab$ alignment [80, 81]. However, the precise assignment of the $H_0 \parallel ab$ singularity in powder spectra is ambiguous.

The Knight shift of sample S2 is slightly larger than the one of S1. More strikingly, it shows a clear decrease below $T_c$ for both measured angles ($\theta = 0°$ and $3°$), and does not exhibit a strong angle dependence.

The constant Knight shift for sample S1 and $\theta = 0°$ suggests the occurrence of unusual triplet superconductivity in this sample. But before really concluding on triplet Cooper pairing, several other possible sources for a constant Knight shift in the superconducting state have to be ruled out.

The total Knight shift $K$ consists of an orbital and a spin contribution: $K = K_{orb} + K_s$. The orbital part involves the orbital motion of the conduction electrons and is usually temperature-independent, while the spin shift $K_s$ is proportional to the static spin susceptibility $K_s \propto A_{hf}(\vec{q} = 0, \omega_L)\chi_s(\vec{q} = 0)$. As already stated in Section 3.1, in the case of strong orbital magnetism, the orbital shift $K_{orb}$ can become as large as the spin shift $K_s$ and may lead to an unchanged Knight shift across $T_c$ even for singlet superconductors [92, 93]. Also spin-orbit scattering in the presence of disorder might cause a finite Knight shift in the superconducting state [94]. However, both effects can be ruled out for the present case. First of all, the strong angular dependence of the Knight shift in the superconducting state (see left panel of Fig. 8.5) is incompatible with a large orbital contribution to the Knight shift. Second, the vast collection of proofs for the high purity of the crystals [small NQR and NMR lines, sharp superconducting transition, smallest residual resistivity, ... (see Section 5.3.2)] speaks strongly against the scenario of an impurity-induced enhanced spin-orbit scattering.

Diamagnetic shifts due to demagnetization effects in the superconducting state always lead to a decreasing shift (see Section 3.1). Together with an increasing spin shift, this could also result in the observation of an unchanged total shift in the superconducting state. However, due to the plate-like shape of the samples diamagnetic shifts can be neglected for $H_0 \parallel ab$. Furthermore, large demagnetization effects are normally also visible in a line broadening due to an inhomogeneous field distribution in the mixed state. Such a line broadening is not observed for sample S1. In contrast, the NQR and NMR linewidth decreases in the superconducting state of sample S1.

After ruling out all the mentioned possible sources for a constant Knight shift, one can conclude that the observation of a constant Knight shift in the superconducting state of sample S1 for $H_0 \parallel ab$ is caused by an unchanged spin susceptibility. This gives strong evidence for an unconventional spin-triplet superconducting pairing state in sample S1.

The stark changes in the temperature dependence of the Knight shift in the superconducting state upon slightly tilting the sample S1 in the magnetic field suggest that even small magnetic fields along $c$ direction are enough to cause a transition into another superconducting state, with a different and possibly more conventional spin-singlet symmetry.

In triplet superconductors, a strong anisotropy of the spin susceptibility is anticipated because of spin-orbit coupling, which, even when it is weak, will align the spins of the triplet pairs along a particular direction. The observed anisotropy of the Knight shift (and thus the spin susceptibility) in LiFeAs indicates that the spins of the Cooper pairs lie within the $ab$ plane. The $\vec{d}$-vector, which characterizes the order parameter of spin-triplet superconductors (see Section 3.1), thus points along $c$. A similar situation is encountered in $Sr_2RuO_4$. There, the phase with the order parameter $\vec{d} \parallel c$ is stable as long as the field is applied along $ab$. Already a small tilting angle results in drastic changes in the upper critical field $H_{c2}$ which were interpreted as a transition to another superconducting state [302].

The right panel of Fig. 8.5 shows the Knight shift of sample S3 at low temperature, measured for $H_0 \parallel ab$ for fields of 7 T and 4.5 T. Because of possible heating effects which might falsify the quadrupole frequency at 4.2 K (see Section 5.4.3), only data for temperatures down to 8 K can be shown.[2] The first measurements in 7 T (blue dots) showed an unchanged Knight shift across $T_c$, similar to what has been observed for sample S1, and thus also points towards a spin-triplet pairing mechanism in this sample. Subsequent measurements at lower field values of 4.5 T (orange dots) showed a decrease of the Knight shift in the superconducting state. The first assumption at this stage of the measurements was that the sample was aged and thus degraded. However, a subsequent check at 7 T (dark red dots) showed within error bars the same absolute values of $K(H_0 = 7\,\mathrm{T})$ as obtained in the measurements at 7 T before and again a constant Knight shift across $T_c$.

[2] Heating effects in the NMR measurements can be excluded, since all data sets, displaying constant and decreasing Knight shifts, have been measured under the same measurement conditions.

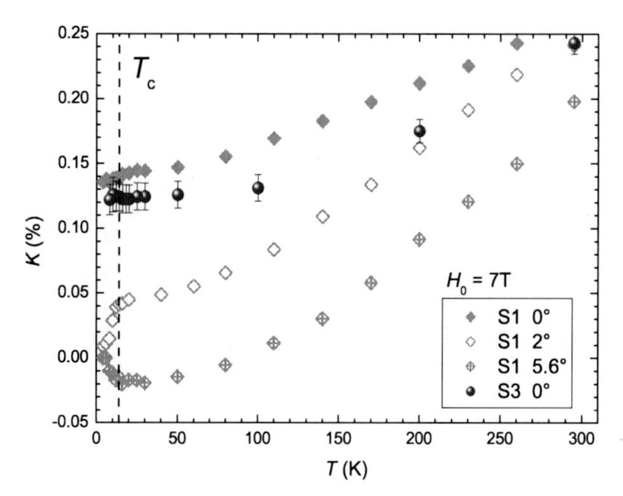


**Figure 8.6:** $^{75}$As NMR Knight shift of LiFeAs samples S1 (filled grey rhombi) and S3 (blue dots) for $H_0 \parallel ab$ in $H_0 = 7\,\mathrm{T}$ in the whole temperature range. For sample S1, additional data for small tilting angles of the magnetic field vs the *ab* direction are shown (open grey rhombi). The dashed line denotes $T_c(7\mathrm{T}) = 14.6\,\mathrm{K}$.

Besides the strong angular dependence of the Knight shift as revealed by measurements on sample S1, LiFeAs also seems to feature a strong field dependence of its superconducting properties. The observation of constant Knight shift in samples S1 and S3 at a magnetic field of 7 T and the decreasing Knight shift of sample S3 in a lower field is reminiscent of the situation found in one-dimensional superconductors, where a field-induced transition from singlet to triplet superconductivity has been discussed theoretically and experimentally [303, 304].

It is clear that more data are needed to clarify the situation encountered in LiFeAs. Unfortunately the high sensitivity of the samples to moisture and air precluded measurements of the angle-dependent Knight shift of sample S3 and/or the field-dependent Knight shift of sample S1.

To complete the Knight shift data set, Fig. 8.6 shows the Knight shift of the samples S1 and S3 in the paramagnetic phase. In agreement with earlier NMR work on powder samples [80, 81], it decreases with decreasing temperature before it levels off at around 50 K. Analogous to the procedure in Section 6.1, the hyperfine coupling constant was extracted from plots of $K$ versus the bulk susceptibility for $T > 170\,\mathrm{K}$. For $H \parallel ab$, the averaged hyperfine coupling amounts to $^{75}A_{hf}^{ab} = 63\,\mathrm{kOe}/\mu_B$ [298]. This value is enhanced compared to the hyperfine coupling of $^{75}$As in $LaO_{0.9}F_{0.1}FeAs$ (see Table 6.1). An extraction of the hyperfine coupling constant for $H \parallel c$ gave $^{75}A_{hf}^{c} = 9.3\,\mathrm{kOe}/\mu_B$ [298] and thus a large anisotropy of the hyperfine couplings, in contrast to what has been observed for other pnictides (for instance $^{75}A_{hf}^{ab} = 26.4\,\text{kOe}/\mu_B$ and $^{75}A_{hf}^{c} = 18.8\,\text{kOe}/\mu_B$ in the case of $BaFe_2As_2$ [228]).

The flattening at around 50 K evidences that the static spin susceptibility is enhanced at low temperatures. This is consistent with the presence of ferromagnetic spin fluctuations, which were suggested by theoretical calculations [214]. A similar behavior has been observed in $Sr_2RuO_4$ [233]. The decrease of $K$ at higher temperatures resembles the decreasing Knight shift observed in many other pnictides. It may have a different origin and might even mask the enhancement at low temperature. The strong angular dependence of $K$ measured for sample S1 is difficult to understand. It may be due to the complicated multi-band character of the electronic structure. In such a case, the spin susceptibility may have different contributions from different bands, which might respond differently to a given field direction. This is similar to the case of $Sr_2RuO_4$, where an orbital-dependent behavior of the Ru $4d$ spin susceptibility has been observed and related to several independent spin degrees of freedom in this compound [233].

## 8.2.2 Dynamic Properties - Spin-Lattice Relaxation Rate

Fig. 8.7 compares the $^{75}$As NMR spin-lattice relaxation rates for all three single crystals measured along $H_0 \parallel ab$. In the normal state at high temperature, samples S1 and S3 exhibit comparable absolute values of $(T_1T)^{-1}$. Sample S3 follows a similar behavior as observed in $^{75}$As NQR $(T_1T)^{-1}$: it decrease slightly with temperature before levelling off at a constant value at around 120 K. In contrast, $(T_1T)^{-1}$ of sample S1 shows nearly no temperature dependence at high temperature and increases to a weak but clearly visible maximum just above $T_c$. This increase takes place in the temperature range where $(T_1T)^{-1}$ of S3 is constant. The observed peak is also visible in the $(T_1T)^{-1}$ data of the same sample measured along $c$ direction (see left panel of Fig. 8.8). Its presence in the $^{75}$As NMR $(T_1T)^{-1}$, together with its absence in the $^{75}$As NQR $(T_1T)^{-1}$ of sample S1, indicates that the normal state spin dynamics are strongly influenced by an external magnetic field, signaling a proximity to an instability, whose nature is still unclear.

The behavior of $(T_1T)^{-1}$ of sample S1 also remains unusual in the superconducting state. It shows only slight changes upon crossing $T_c$ (see inset of Figure 8.7). The strong upturn of $(T_1T)^{-1}$ below $T_c$, which was observed in $^{75}$As NQR measurements on S1 (see Fig. 8.3), is suppressed in field. $(T_1T)^{-1}$ of S2 clearly drops below $T_c$, similarly to what has been reported in previous works on powder and single crystalline samples [80, 81, 225]. For sample S3, $(T_1T)^{-1}$ first shows a slight enhancement directly below $T_c$, which is then followed by a decrease. This has also been observed in $^{75}$As NQR $(T_1T)^{-1}$ measurements on this sample (see inset of Fig. 8.3). In the NQR measurements it seems that $(T_1T)^{-1}$ of S3 first follows the enhancement of $(T_1T)^{-1}$ of S1 and then decreases. Since this increase takes place below $T_c$, it possibly indicates the presence of spin fluctuations in the superconducting state of sample S3. Note that such an increase has not been observed for sample S2.

To further compare the upturn in the $^{75}$As NQR $(T_1T)^{-1}$ to the suppression of this upturn in a magnetic field, the left panel of Fig. 8.8 compares the $^{75}$As NQR $(T_1T)^{-1}$ of S1 with the $^{75}$As NMR $(T_1T)^{-1}$ of S1 measured along $H_0 \parallel c$ at low temperature. Since $c$ is the principal axis of the EFG, both data sets should be equivalent. And indeed the absolute values are comparable at high temperature. At low temperature, however,

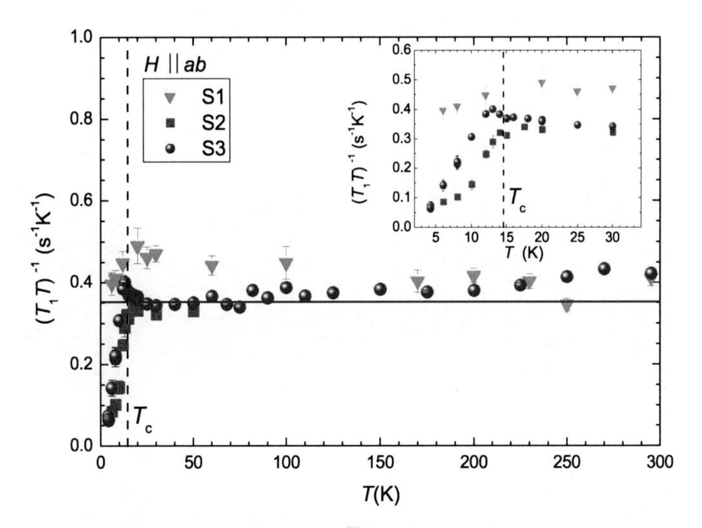


**Figure 8.7:** Temperature dependence of the $^{75}$As NMR spin-lattice relaxation rate measured along $H_0 \parallel ab$ for samples S1 (grey triangles), S2 (red squares) and S3 (blue dots). The blue line is a guide to the eye. The inset enlarges the low temperature part. The dashed lines denote $T_c(H_0)$. The data were taken in $H_0 = 7\,\mathrm{T}$ for S1 and S3 and in $H_0 = 8.5\,\mathrm{T}$ for S2.

both data sets deviate from each other. While the $^{75}$As NQR $(T_1T)^{-1}$ is constant just before $T_c$, the $^{75}$As NMR $(T_1T)^{-1}$ shows the already mentioned peak just above $T_c$. In the superconducting state, the $^{75}$As NQR $(T_1T)^{-1}$ increases, while the $^{75}$As NMR $(T_1T)^{-1}$ does not exhibit strong changes.

A similar behavior is observed in the superconducting state of the polycrystalline sample (see right panel of Fig. 8.3). Its $^{75}$As NQR $(T_1T)^{-1}$ shows a strong upturn in the superconducting state. The beginning of this upturn is clearly connected to $T_c$ and thus to the superconducting state. In the $^{75}$As NMR $(T_1T)^{-1}$ measured in a field of 15.47 T parallel to $ab$ [$T_c(15.47\,\mathrm{T}) \approx 12\,\mathrm{K}$], this upturn is completely suppressed and $(T_1T)^{-1}$ decreases very weakly below $T_c$.

A very similar behavior has also been observed in $La_{0.87}Ca_{0.13}OFeP$ by $^{31}$P NMR measurements in different fields [297]. The data are reproduced in Fig. 8.9. In this compound, the $^{31}$P NMR $(T_1T)^{-1}$ increases below $T_c$ as long as very small external fields are applied. This increase is suppressed by applying larger fields.

In the following, the arguments of Reference [297] will be repeated, which led to the conclusion that novel spin dynamics develop below $T_c$. At first glance, one might relate the increase of $(T_1T)^{-1}$ in the superconducting state with some impurity contribution which may dominate the spin dynamics in the superconducting state as other contributions to the dynamics are gapped out. However, as already pointed out various times, the high purity reflected in sharp $^{75}$As NMR, $^{75}$As NQR and $^{7}$Li NMR resonance lines, as well as other physical properties and also the possibility to fit the recovery of the nuclear

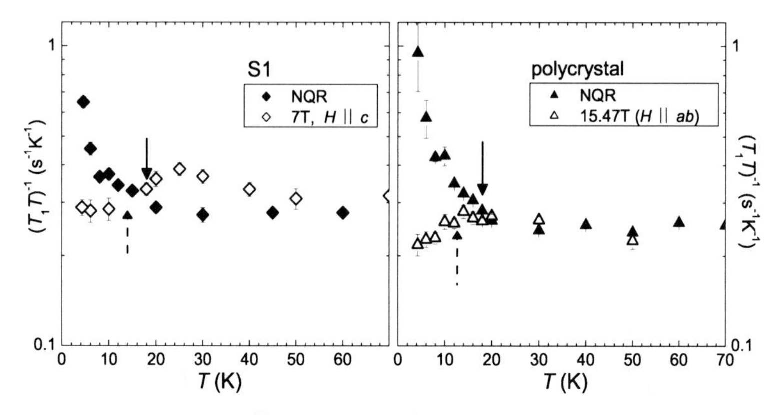


**Figure 8.8:** Comparison of the $^{75}$As NQR $(T_1T)^{-1}$ (closed symbols) and the $^{75}$As NMR $(T_1T)^{-1}$ (open symbols) for S1 (rhombi, left panel) and the polycrystalline sample (triangles, right panel). Note that for S1, the $^{75}$As MNR $(T_1T)^{-1}$ has been measured in a field of 7 T and parallel to $c$, while for the polycrystal it has been measured in a field of 15.47 T and along $ab$. The $^{75}$As NMR $(T_1T)^{-1}$ data of the polycrystal are scaled to facilitate comparison with the NQR data, which, since $V_{ZZ} \parallel c$, measure along $c$. Due to an anisotropy in the hyperfine coupling $A_{hf}(\vec{q})$, the actual absolute value of the $^{75}$As NMR $(T_1T)^{-1}$ for $H \parallel ab$ is higher. The arrows denote $T_c$ with (dashed) and without (solid) external field.

magnetization with a single exponential function in the whole temperature range, makes an impurity contribution very unlikely. An increase of $(T_1T)^{-1}$ due to vortex dynamics, which sometimes has been observed in organic superconductors and cuprates [305, 306], can also be excluded since the upturn of $(T_1T)^{-1}$ in LiFeAs is most pronounced in $^{75}$As NQR measurements, where vortices are absent. A third possibility considers the existence of local moments in the system. In the normal state these local moments scatter with the conduction electrons and fluctuate very fast, such that they do not affect the NMR properties. Upon entering the superconducting state, the conduction electrons are gapped out. Therefore, the local moments start to slow down, leading to an increase in the spin-lattice relaxation rate. However, if local moments would be present, they should also affect the linewidth and the Knight shift, which is not the case. So also this possibility is unlikely.

The last possibility pointed out by Nakai *et al.* [297], is the emergence of low-energy spin fluctuations relevant to the superconducting order parameter. In the case of spin-triplet superconductivity, the collective modes of the Cooper pairs may give rise to novel spin dynamics in the superconducting state [297]. This explanation, independently suggested when discussing the $(T_1T)^{-1}$ data of $La_{0.87}Ca_{0.13}OFeP$,[3] is in good agreement with the observation of a constant Knight shift across $T_c$ for the same sample S1, for which the enhancement of $(T_1T)^{-1}$ in the $^{75}$As NQR was observed. Thus, there are two independent strong evidences for the occurrence of spin-triplet superconductivity in sample S1.

[3] Note that up to now there are no published Knight shift measurements on $La_{0.87}Ca_{0.13}OFeP$.

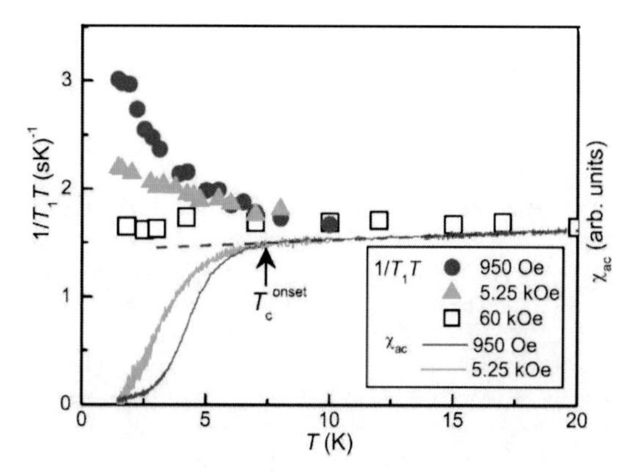


**Figure 8.9:** $^{31}$P NMR $(T_1T)^{-1}$ of $La_{0.87}Ca_{0.13}OFeP$ at low temperature, measured in different external fields of 0.095 T (blue dots), 0.525 T (green triangles) and 6 T (open squares). The solid lines denote the ac susceptibility $\chi_{ac}$ in 0.095 T (blue) and 0.525 T (green). The red dashed line is a linear fit to $\chi_{ac}$ to define $T_c$. Figure reproduced from [297].

The $^{75}$As NQR and NMR spin-lattice relaxation rate $T_1^{-1}$ of sample S3 is plotted in Fig. 8.10. It decreases below $T_c$ with a field-independent $T^3$ dependence. This much more usual behavior is surprising, since the Knight shift measured in 7 T did not change upon crossing $T_c$, pointing towards a spin-triplet superconducting state similar to the Knight

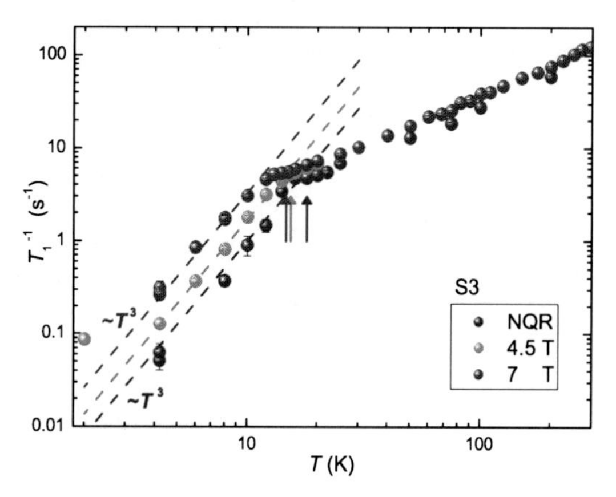


**Figure 8.10:** Temperature dependence of the $^{75}$As NQR $T_1^{-1}$ (blue squares) and $^{75}$As NMR $T_1^{-1}$ in 4.5 T (orange dots) and 7 T (violet dots) of sample S3 on a doubled logarithmic scale. Arrows in the respective colour denote the corresponding $T_c$ (18 K for NQR, 15.3 K for 4.5 T and 14.6 K for 7 T). The dashed lines represent $T_1^{-1} \propto T^3$ in the superconducting state.

shift of S1. Regarding these very similar static properties between these two compounds one could have expected that S3 also features the same dynamic properties than S1. This is clearly not the case. The reason for this discrepancy is unclear.

## 8.3 Summary of $^{75}$As NQR and NMR Results

Taken together all the obtained $^{75}$As NQR and NMR measurements on the three different samples out of the same batch, no definite picture of the superconducting properties of LiFeAs can be given at the end. For the three investigated single crystals and the polycrystalline sample, different normal state properties as well as different superconducting state properties have been observed. They are summarized in Table 8.1. The single crystalline samples exhibit a high purity, confirmed by different physical properties as well as by several NMR and NQR aspects such as narrow NQR and NMR resonance lines and single exponential recovery curves. The samples are only distinguishable by their slightly different quadrupole frequencies $\nu_q$, suggesting tiny changes in their stoichiometries which cannot be measured by any macroscopic measurement. These tiny changes are enough to cause strongly different superconducting properties.

For sample S1, the sample with the highest $\nu_q$, strong evidences for a spin-triplet superconducting state (as long as the field is applied exactly along *ab*) are found in the constant Knight shift and the peculiar behavior of $(T_1T)^{-1}$. The strong angular dependence of the static properties suggests that a transition into another superconducting state with a more usual order parameter may be caused by small magnetic fields along the *c* direction.

Sample S2, the one with the lowest $\nu_q$, exhibits the "least surprising" properties. Its Knight shift as well as its $^{75}$As NMR $(T_1T)^{-1}$ decrease in the superconducting state. It also lacks a strong angular dependence. This is compatible with standard singlet pairing.

Sample S3 exhibits the intermediate $\nu_q$ and also lies in between the two other samples regarding its properties. For NQR and NMR measurements, its $(T_1T)^{-1}$ first increases slightly just below $T_c$, before decreasing similarly to $(T_1T)^{-1}$ of S2. This may indicate the persistence of spin fluctuations in the beginning of the superconducting state. The Knight shift of S3 is constant across $T_c$ when measured in 7 T, but decreases when measured in 4.5 T. This points towards a field-induced triplet superconducting state similar to the one observed in S1.

The superconducting state of LiFeAs is thus believed to be strongly dependent on the stoichiometry, the absolute value of the magnetic field and the field-orientation. Different phases are competing in LiFeAs and tiny changes of the stoichiometry, the field or the angle can cause transitions between different superconducting ground states, including the possibility of spin-triplet pairing. The extreme sensitivity of the properties towards tiny changes of external parameters suggests that LiFeAs is in a proximity to an instability. As suggested by recent theoretical work [214] and corroborated by the evidences for spin-triplet pairing in some cases, this instability might be a ferromagnetic one.

The overall behavior of LiFeAs is clearly unconventional and differs a lot from other pnictides. A lot of questions remain unanswered. Further investigation of the angular and field dependence are needed to clarify the situation encountered in this interesting material.

| property | S1 | poly | S3 | S2 |
|---|---|---|---|---|
| $\nu_q(100\,\mathrm{K})$ (MHz) | 21.19 | 21.18 | 21.15 | 21.13 |
| $\mathrm{FWHM_{NQR}}(300\,\mathrm{K})$ (kHz) | 80 | 113 | 44 | 60 |
| $\mathrm{FWHM_{NQR}}$ in normal state | increases | - | increases | const. |
| $\mathrm{FWHM_{NQR}}$ in SC state | decreases | increases (not shown) | increases | - |
| $(T_1T)^{-1}_{\mathrm{NQR}}$ in SC state | increases | increases | decreases | - |
| $\mathrm{FHWM_{NMR}}(20\,\mathrm{K})$ (kHz) | 52 | - | 38 | 17 |
| $\mathrm{FWHM_{NMR}}$ in normal state | increases | - | increases | const. |
| $\mathrm{FWHM_{NMR}}$ in SC state | decreases | - | both | increases |
| $K_{ab}(7\,\mathrm{T})$ in SC state | const. | - | const. | decreases |
| $K(7\,\mathrm{T},\,\theta > 0°)$ in SC state | decreases | - | ? | decreases |
| $K_{ab}(4.5\,\mathrm{T})$ in SC state | ? | - | decreases | ? |
| $(T_1T)^{-1}_{\mathrm{NMR}}$ in SC state | approx. const. | decreases slightly | decreases | decreases |

**Table 8.1:** Summary of some NQR and NMR properties of all four investigated samples of LiFeAs. They are ordered according to their quadrupole frequency (see second row).

## 8.4 $^7$Li-NMR

$^7$Li NMR measurements were performed on sample S3 in a field of 4.5 T for both field directions, $H_0 \parallel c$ and $H_0 \parallel ab$. Some selected spectra for both directions are presented in Fig. 8.11. At high temperature very narrow spectra are observed for both directions, with the same linewidths for the central line and the satellites. At $T = 200\,\mathrm{K}$ the linewidths amount only to 11 kHz (9 kHz) for $H_0 \parallel c$ ($H_0 \parallel ab$), respectively. These small linewidths allow to observe the very small splitting of the $^7$Li NMR spectra into a central line and two satellites due to first order quadrupole effects, although the quadrupole frequency is only 32 kHz at room temperature. The satellites for $H_0 \parallel c$ are separated by $2\nu_q$ and the satellites for $H_0 \parallel ab$ are separated by $\nu_q$. According to Fig. 2.2 this corresponds to a tetragonal symmetry with an asymmetry parameter $\eta = 0$ and the principal axis of the EFG lying along the $c$ axis.

The identical linewidth of the central line and satellites for a given field direction at high temperatures indicates that the broadening mechanism is mainly magnetic and the distribution of the EFG is negligibly small. The extreme narrow lines corroborate further the high purity of the investigated sample. Furthermore, the $^7$Li spectra assure that there is only one single Li site in the investigated sample. This is in stark contrast to what has been observed by $^7$Li NMR measurements on a single crystal by another group, where two different $^7$Li resonances are observed, pointing towards the presence of two inequivalent sites in their crystal [225].

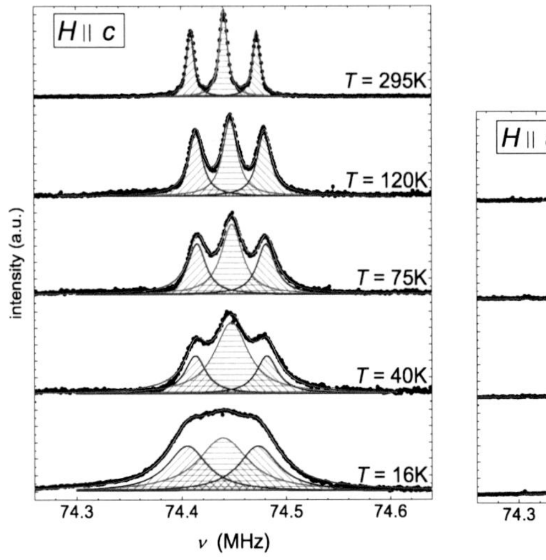


(a) $^{7}$Li NMR spectra for $H_0 \parallel c$. The satellites are separated by $2\nu_q$.

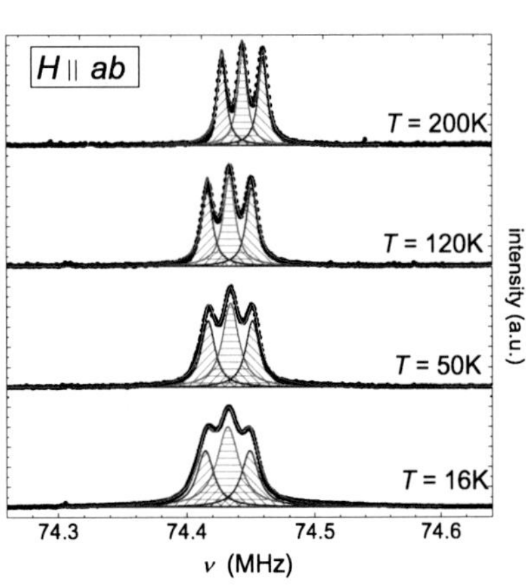


(b) $^{7}$Li NMR spectra for $H_0 \parallel ab$. The satellites are separated by $\nu_q$.

**Figure 8.11:** $^{7}$Li NMR spectra of LiFeAs for both field directions at some selected temperatures. The black dots denote the measured spectra, the red line is a fit containing three Lorentzian lines connected via the Knight shift $K$ and the quadrupole frequency $\nu_q$ and with the constraints, that both satellites feature the same linewidth and area. The green (blue) filled lines denote the corresponding Lorentzian line(s) for the central line (satellites).

By lowering the temperature the resonance lines broaden. It is possible to fit the spectra with three Lorentzian lines connected by $K$ and $\nu_q$ and with equal linewidths of the satellites down to $T_c$ for both field directions. Fig. 8.12 comprises the temperature evolution of the linewidths of the satellites and the central lines for $H_0 \parallel ab$ and $H_0 \parallel c$. The broadening is more pronounced for $H_0 \parallel c$. The broadening in the normal state indicates that the $^{7}$Li nuclei experience a slowing down of spin fluctuations, similar to the $^{75}$As nuclei (see Fig. 8.4).

Below $T_c$, the spectra for $H_0 \parallel ab$ suddenly broaden much more, such that a fit with three Lorentzians is not possible any more. On this account, only data down to 14 K are shown for $H_0 \parallel ab$. The values of $K$ and $\nu_q$ could also not be extracted consistently below 14 K for $H_0 \parallel ab$. In the case of $H_0 \parallel c$ the spectra below $T_c$ resemble the one for 16 K shown in Fig. 8.11(a) and $K$ and $\nu_q$ could still be extracted, although with larger error bars.

The value of the extracted quadrupole frequency, $\nu_q(300\,\mathrm{K}) = 32\,\mathrm{kHz}$, is consistent with the value of $\nu_q(300\,\mathrm{K}) = 34\,\mathrm{kHz}$, which was estimated by an echo decay measurement on a polycrystalline sample [80]. The temperature evolution of the quadrupole frequency is depicted in Fig. 8.13. It shows a much weaker temperature dependence than the $^{75}$As quadrupole frequency. No unusual contributions to the EFG and its temperature

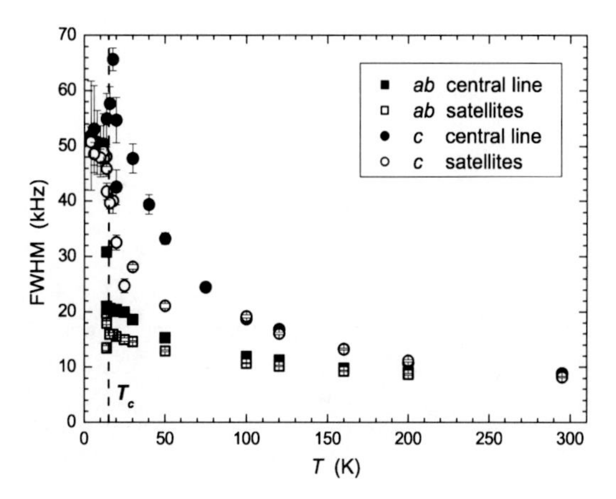


**Figure 8.12:** Temperature dependence of the full width at half maximum (FWHM) of the $^7$Li NMR resonance lines for $H_0 \parallel ab$ (squares) and $H_0 \parallel c$ (dots). The FWHM for the central lines (closed symbols) are the same as the ones for the satellites (open symbols) at high temperature. The dashed line denotes $T_c(4.5\,\mathrm{T}) = 15.3\,\mathrm{K}$.

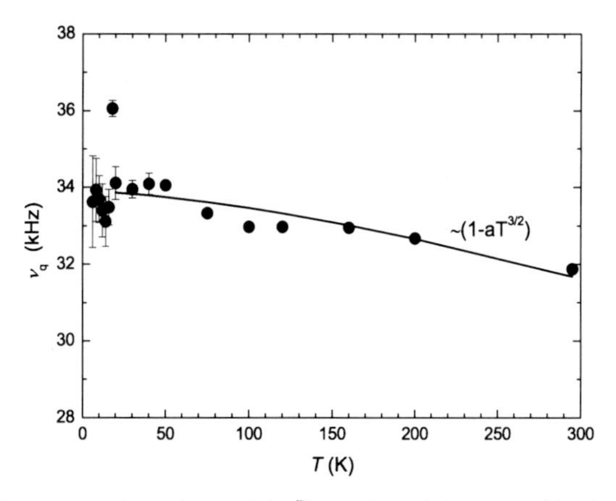


**Figure 8.13:** Temperature dependence of the $^7$Li quadrupole frequency (dots). The solid line is a fit to the empirical temperature dependence of the quadrupole frequency in non-cubic metals [307].

dependence are visible. Its slight increase with decreasing temperature is in line with the frequently observed behavior of the EFG in non-cubic metals, showing a $(1 - aT^{3/2})$ dependence [307]. At low temperature, error bars increase due to the broadening of the spectra.

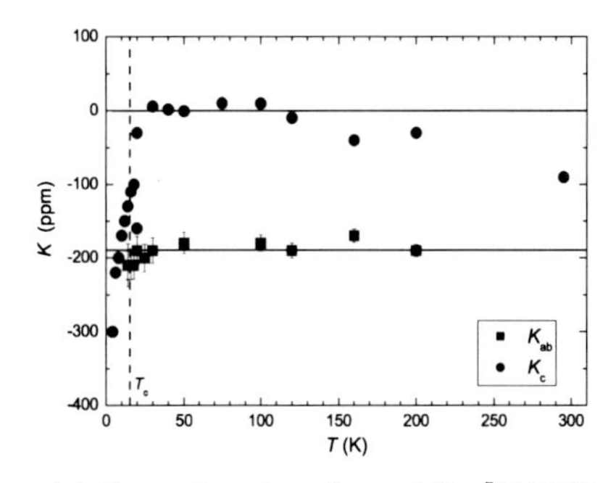


(a) Temperature dependence of the $^7$Li NMR Knight shift for $H_0 \parallel ab$ (squares) and $H_0 \parallel c$ (dots) measured in $H_0 = 4.5\,\mathrm{T}$. The dashed line denotes $T_c(4.5\,\mathrm{T}) = 15.3\,\mathrm{K}$. The solids lines are guides to the eyes.

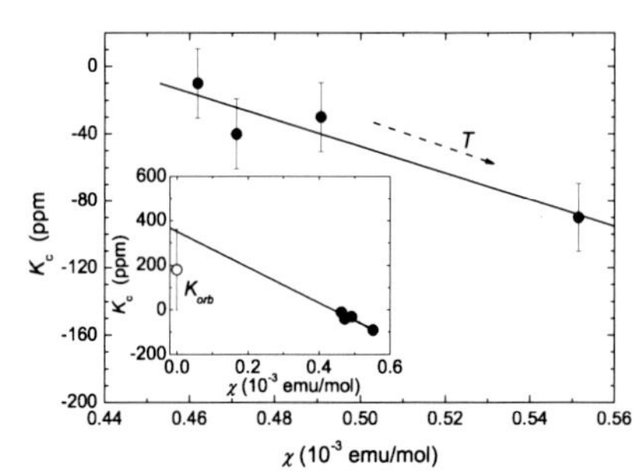


(b) $^7$Li NMR Knight shift for $H_0 \parallel c$ vs macroscopic susceptibility $\chi$. The inset shows the extraction of $K_{orb}$.

**Figure 8.14:** Temperature-dependent $^7$Li NMR Knight shift for both field directions (*a*) and extraction of the hyperfine coupling tensor for $H_0 \parallel c$ (*b*).

The Knight shift for $H_0 \parallel ab$ and $H_0 \parallel c$ is depicted in Fig. 8.14(a). Again, it can be divided into a temperature-independent orbital contribution $K_{orb}$ and the spin part $K_s$, which stems from a hyperfine coupling to the electronic spin susceptibility. A pure orbital shift for Li in solids is typically found to range in between $-20\,\mathrm{ppm}$ and $+30\,\mathrm{ppm}$ [308]. The measured Knight shift is very small and negative. For $H_0 \parallel c$ it is basically zero at intermediate temperatures. At high temperatures it slightly increases. Since the macroscopic susceptibility decreases in the high temperature range [129], the increase of the Knight shift in this temperature range indicates that the hyperfine coupling is negative. This is unexpected. Negative contributions to the hyperfine coupling normally stem from core polarization effects (see Section 2.2.2). But since Li is in a $1s^2$ state, which is a fully filled, isotropic shell, no core polarization contributions are expected. Possibly this negative hyperfine coupling stems from a core polarization due to non-$s$-character-like conduction electrons or from an *sp* hybridization with the As $p$ orbitals. A plot of the Knight shift for $H_0 \parallel c$, $K_c$, versus the macroscopic susceptibility $\chi$ with temperature as implicit parameter is shown in Fig. 8.14(b). Only data for $T \geq 120\,\mathrm{K}$ are shown. A linear fit between $K$ and $\chi$ resulted in a hyperfine coupling of about $^7A_{hf}(\vec{q} = 0) = (-4.4 \pm 1.1)\mathrm{kOe}/\mu_B$. This value is comparable to the hyperfine coupling of La in $\mathrm{LaO_{0.9}F_{0.1}FeAs}$ (see Table 6.1). Only the sign is opposite. The relatively small value is consistent with the location of both ions outside the FeAs layers, which leads to a weak coupling to the dominant Fe $3d$ moments. For the orbital contribution only the upper and lower limits can be given, since nothing is known about the absolute values of the temperature-independent contributions to the susceptibility (van Vleck and diamagnetic susceptibility). $K_{orb}$ lies between 350 ppm and -10 ppm, which covers widely the range of the typical orbital shifts. 350 ppm is very unlikely, since this would overcome the absolute value of the Knight shift.

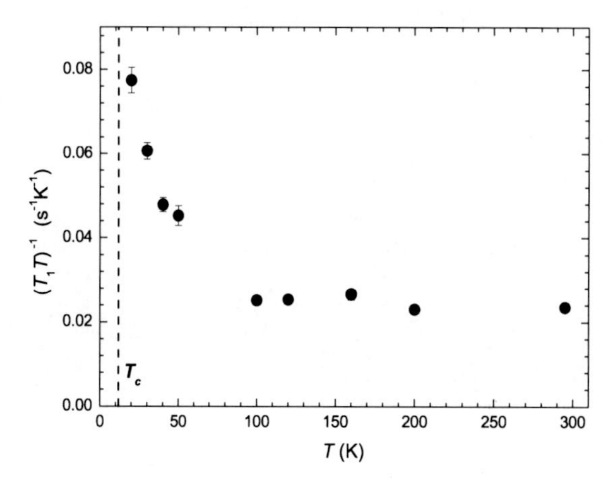


**Figure 8.15:** Temperature-dependent $^7$Li NMR $(T_1T)^{-1}$ for $H_0 \parallel c$ in 4.5 T. The dashed line denotes $T_c(4.5\,\mathrm{T}) = 15.3\,\mathrm{K}$.

For $H_0 \parallel ab$ the Knight shift is constant and amounts to approximately $-190$ ppm. This might be attributed to a pure orbital shift.

The decrease of $K$ for $H_0 \parallel c$ and $T < 30\,\mathrm{K}$ illustrates an increase of the spin susceptibility below 30 K, due to the negative hyperfine coupling constant. Its physical origin is unclear. This increase has to be taken with great care, since in this temperature range the spectra are much broader than at higher temperatures, which renders the determination of the Knight shift more imprecise. An increasing spin susceptibility in the superconducting state is unphysical. It is supposed to decrease in the case of spin-singlet formation or to stay constant in the case of triplet pairing. A possible explanation could be a change of the hyperfine coupling in this temperature range, due to the complicated multi-band character of the electronic structure.

Measurements of $^7$Li NMR spin-lattice relaxation rate were only possible for $H_0 \parallel c$. For $H_0 \parallel ab$ no inversion of the nuclear magnetization could be achieved. Attempts to measure $T_1$ via saturation recovery also failed for $H_0 \parallel ab$. This might be due to the small quadrupole splitting. This small splitting implicates that all three resonance lines are exited during the measurement of $T_1$ (for both field directions). The smaller splitting in the case of $H \parallel ab$ leads to oscillations of the spin echo during the $T_1$ measurements, which complicate the analysis. Fig. 8.15 shows the $^7$Li NMR $(T_1T)^{-1}$ for $H_0 \parallel c$ measured in 4.5 T in the paramagnetic phase. $(T_1T)^{-1}$ is constant down to 100 K and then starts to increase. This increase indicates the slowing down of spin fluctuations, consistent with the increase of the NMR linewidths (see Fig. 8.12). Up to now, no data have been measured in the superconducting state.

## 8.5 Summary of $^7$Li NMR Results

$^7$Li NMR measurements have been performed on sample S3 in a field of 4.5 T. The narrow resonance lines at high temperatures for both field directions corroborate the high purity of the investigated crystal and allow a precise determination of the very small quadrupole frequency ($\nu_q = 32\,\mathrm{kHz}$ at room temperature). A broadening of the resonance lines and an increasing $(T_1T)^{-1}$ with decreasing temperature indicate that the $^7$Li nuclei sense the slowing down of magnetic fluctuations, similar to the $^{75}$As nuclei. The determined Knight shifts are very small and nearly temperature-independent in the paramagnetic regime. A linear scaling between the Knight shift along $c$ direction and the macroscopic susceptibility at high temperature results in a negative hyperfine coupling constant of about $^7A_{hf}(\vec{q} = 0) = (-4.4 \pm 1.1)\mathrm{kOe}/\mu_B$, whose absolute value is comparable to the hyperfine coupling constant of La in $LaO_{0.9}F_{0.1}FeAs$.

# 9 Conclusions

NMR measurements on two compounds of the recently discovered Fe-based superconductors, $LaO_{1-x}F_xFeAs$ and LiFeAs, have been presented. The measurements revealed unconventional properties in the normal state as well as in the superconducting state for both compounds.

**Chapter 6** reported measurements of static and dynamic low-energy properties of $LaO_{1-x}F_xFeAs$ in the normal state, covering a broad doping range from the undoped, magnetically-ordered compound LaOFeAs to the optimally-doped $LaO_{0.9}F_{0.1}FeAs$.

Considering the static NMR properties of $LaO_{1-x}F_xFeAs$, measured by the Knight shift $K$, the following conclusions can be made:

A decrease of the local spin susceptibility with decreasing temperature was found in underdoped $LaO_{0.95}F_{0.05}FeAs$ and optimally-doped $LaO_{0.9}F_{0.1}FeAs$, in agreement with macroscopic susceptibility measurements. The scaling of the local spin susceptibility, measured by the Knight shift, and the macroscopic powder susceptibility proved the absence of magnetic impurity contributions and thereby the high quality of the investigated samples. The further scaling of the NMR Knight shifts of three measured nuclei in $LaO_{0.9}F_{0.1}FeAs$, namely $^{75}$As, $^{139}$La, and $^{57}$Fe, suggested that all nuclear spins are probing the same spin degree of freedom. The system might therefore be describable within a single spin liquid picture, despite the complicated multi-band character of the electronic structure. The decrease of the spin susceptibility in the normal state seems to be a characteristic property of Fe-based superconductors. Its physical reason is still not well understood. Theoretical approaches include effects of a temperature-dependent change in the electronic density of states, and the existence of a pseudogap, similar to what have been found in cuprates. However, no broad pseudogap peak was observed in NMR measurements of the $^{75}$As Knight shift in $LaO_{0.9}F_{0.1}FeAs$ up to 480 K, and intended pseudogap fits result in unrealistically high values of the pseudogap. Together with the dynamic properties, this makes a pseudogap behavior very unlikely and points more towards a density of states effect.

Furthermore, hyperfine coupling constants were extracted for $^{75}$As, $^{139}$La, and $^{57}$Fe nuclei in $LaO_{0.9}F_{0.1}FeAs$, and their absolute values were explained. The ordered moment in the magnetically-ordered SDW state of LaOFeAs was found to be $0.63(1)\,\mu_B$, which is in good agreement with recent neutron scattering experiments.

The investigation of the low-energy spin dynamics by $(T_1T)^{-1}$ measurements in the normal state of $LaO_{1-x}F_xFeAs$ revealed the following results:

For undoped LaOFeAs a strong increase of the $^{75}$As $(T_1T)^{-1}$ towards the magnetic ordering temperature $T_N$ proves the sensitivity of the $^{75}$As nuclei for antiferromagnetic fluctuations. It points towards a second-order-like phase transition to the magnetically-ordered state. The increase of $T_1^{-1}$ can be well described by Moriya's SCR theory for weak itinerant antiferromagnets.

$(T_1T)^{-1}$ of the underdoped samples $LaO_{0.95}F_{0.05}FeAs$ and $LaO_{0.925}F_{0.075}FeAs$ still exhibits a substantial increase with decreasing temperature between $T \approx 200\,K$ and $T_c$, indicating the presence of remnant antiferromagnetic fluctuations in these superconducting compounds. At higher temperatures $(T_1T)^{-1}$ decreases linearly, similar to $(T_1T)^{-1}$ of optimally-doped $LaO_{0.9}F_{0.1}FeAs$. Both the high temperature and the low temperature behavior of $(T_1T)^{-1}$ in the underdoped samples have been compared with the temperature dependence of the resistivity. Good agreement has been found. Resistivity measurements find also evidence for remnant spin fluctuations at low temperatures (above $T_c$) and a non-Fermi-liquid linear temperature dependence at high temperatures.

The dynamic spin susceptibility of $LaO_{0.9}F_{0.1}FeAs$, measured by $(T_1T)^{-1}$, decreases similarly to the static spin susceptibility. No evidence for the existence of antiferromagnetic fluctuations in this optimally-doped sample is found. The decrease of $(T_1T)^{-1}$ with temperature cannot be described within a pseudogap picture. No pseudogap peak is observed in $(T_1T)^{-1}$ up to 480 K and pseudogap fits failed to describe consistently the decrease of $(T_1T)^{-1}$ in the whole measured temperature regime. The decrease can be well described by a simple linear temperature-dependence. A scaling of $(T_1T)^{-1}$ measured on $^{75}As$, $^{139}La$, $^{57}Fe$, and $^{19}F$ in optimally-doped $LaO_{0.9}F_{0.1}FeAs$ suggested that spin fluctuations are suppressed simultaneously over the whole $\vec{q}$-space.

This overall doping dependence of $(T_1T)^{-1}$ in $LaO_{1-x}F_xFeAs$ indicates that antiferromagnetic spin fluctuations, which are still present in the underdoped samples $LaO_{0.95}F_{0.05}FeAs$ and $LaO_{0.925}F_{0.075}FeAs$, have to be suppressed completely before superconductivity can reach its highest $T_c$ in optimally-doped $LaO_{0.9}F_{0.1}FeAs$. In conclusion, superconductivity and magnetism seem to compete in $LaO_{1-x}F_xFeAs$, in contrast to other pnictides such as $Ba(Fe_{1-x}Co_x)_2As_2$, where samples with the highest $T_c$ still feature well pronounced antiferromagnetic fluctuations. These results are in agreement with the different phase diagrams reported for both compounds, where coexistence between magnetism and superconductivity is found for $Ba(Fe_{1-x}Co_x)_2As_2$, but not for $LaO_{1-x}F_xFeAs$.

For $LaO_{0.9}F_{0.1}FeAs$ a Korringa ratio $T_1TK_s^2 = \text{const.}$, pointing towards a Fermi-liquid behavior, was observed for $T_c < T < 300\,K$ in good agreement with resistivity measurements. A more unconventional linear scaling $T_1TK_s = \text{const.}$ was observed at higher temperatures for $LaO_{0.9}F_{0.1}FeAs$ and for $T \geq 250\,K$ for $LaO_{0.95}F_{0.05}FeAs$. The slope of both linear scalings was found to be the same. This suggests an intrinsic, doping-independent high temperature electronic phase persisting over a broad doping range. Its microscopic origin is still not well understood, but might be closely related to the complicated multi-band electronic structure.

The properties of $LaO_{1-x}F_xFeAs$ in the superconducting state were discussed in **Chapter 7**.

The $^{75}As$ NMR and $^{75}As$ NQR spin-lattice relaxation rate $T_1^{-1}$ was found to decrease proportional to $T^3$ in the superconducting state of $LaO_{1-x}F_xFeAs$ with $x = 0.05$, $x = 0.075$ and $x = 0.1$. Together with the absence of a Hebel-Slichter coherence peak this points towards an unconventional superconducting gap symmetry. Taking into account the results of other experimental methods which found evidence for a fully-gapped superconducting order parameter, the results indicated an unconventional $s_\pm$ symmetry of the superconducting gap function. Within this symmetry, each Fermi surface features an isotropic superconducting gap, whose sign changes upon going from a hole to an electron pocket, and vice versa. The observed temperature dependence of the spin-lattice relaxation rate can be well described within this model by considering the effect of impurity scattering or weak anisotropies in the gap function.

The effect of impurities on the superconducting properties was further studied on an optimally-doped sample with additional arsenic vacancies, $LaO_{0.9}F_{0.1}FeAs_{1-\delta}$. Despite the enhanced disorder, this compound possesses improved superconducting properties and exhibits a faster decrease of the spin-lattice relaxation rate in the superconducting state. $T_1^{-1}$ decreases with $T^5$. The effect of arsenic impurities was first discussed under the assumption that these vacancies act as non-magnetic impurities. Three possible scenarios were presented under this assumption: a crossover from a nodal to nodeless $s_\pm$ symmetry, an impurity-induced enhanced intraband scattering, which stabilizes the $s_\pm$ symmetry, and a transition from a $s_\pm$ to a more conventional $s_{++}$ symmetry. Since recent analyses of the spin susceptibility provided evidence for the formation of local moments around each arsenic vacancy, the spin-lattice relaxation rate data were reinterpreted by considering the effect of magnetic impurities. In this case, the $s_\pm$ symmetry is stabilized, because magnetic impurities are harmless for such a sign-reversing symmetry.

Further work might concentrate on the overdoped region, to test the Fermi-liquid behavior as suggested by resistivity measurements for $x > 0.1$. Also the impact of impurities might be further studied, for instance by changing the concentration of arsenic deficiencies in $LaO_{0.9}F_{0.1}FeAs$ or by examining their effects in underdoped samples.

**Chapter 8** reported $^{75}$As NQR, $^{75}$As NMR and $^{7}$Li NMR measurements on LiFeAs.

$^{75}$As NQR and NMR measurements on three different single crystals and a polycrystalline sample of LiFeAs yielded very different results in the normal state as well as in the superconducting state. The samples could be distinguished by their slightly different $^{75}$As NQR frequencies, indicating tiny changes in the stoichiometry, which were not detectable by other experimental methods up to date.

The sample with the highest $\nu_q$ (S1) exhibits an unchanged Knight shift $K_{ab}$ upon crossing $T_c$ as long as the magnetic field is exactly aligned along the *ab* direction. Furthermore a peculiar upturn of the $^{75}$As NQR $(T_1T)^{-1}$ in the superconducting state is observed for this sample, which is suppressed in $^{75}$As NMR measurements. These results indicate the possibility of a spin-triplet superconducting ground state in sample S1. The angle dependence of $K$ suggests that a transition into another superconducting symmetry with a more usual singlet pairing occurs as soon as the magnetic field is slightly tilted out of the *ab* plane.

The crystal with the lowest $\nu_q$ (S2) exhibits more usual NMR properties, such as a decrease of the Knight shift and the spin-lattice relaxation rate in the superconducting state, both being compatible with unconventional singlet pairing.

Sample S3 (with an intermediate $\nu_q$) lies between these two extremities. $(T_1T)^{-1}$ of this sample decreases in the superconducting state for both, $^{75}$As NQR and $^{75}$As NMR measurements. Its Knight shift decreases in low fields (4.5 T), but is constant in slightly higher fields (7 T) in the superconducting state.

The superconducting state of LiFeAs is thus found to be strongly dependent on slight changes of the stoichiometry, as well as on the value and the relative orientation of the magnetic field.

The samples with the most unusual properties in the superconducting state (S1 and S3), show an increase of the $^{75}$As NQR and $^{75}$As NMR linewidths with decreasing temperature, indicating the presence of spin fluctuations in these compounds. Recent theoretical work suggests a proximity of this compound to a ferromagnetic instability.

$^{7}$Li NMR measurements on sample S3 in the paramagnetic regime showed a weak temperature dependence of the quadrupole frequency and a vanishingly small, nearly temperature-independent Knight shift. An increase of the spin-lattice relaxation rate below 100 K proved that also $^{7}$Li is sensitive to the slowing-down of magnetic fluctuations.

In conclusion, the situation in LiFeAs is far from being well understood. The supposed slight differences in stoichiometry should be checked with other high-sensitive methods. More data for different field values and field orientations are needed to understand the sensitivity of the ground state.

Ongoing work in the IFW Dresden also concentrates on NMR measurements on ferromagnetic Li-deficient samples and the effect of Co-doping.

# A Appendix

## A.1 NMR Powder Spectra

According to Equations (2.33) and (2.34), the first and second order quadrupole shifts depend on the Euler angles $\theta$ and $\phi$, which describe the relative orientation of the principle axis system of the EFG $(X, Y, Z)$ to the magnetic field direction (commonly along $z$). The principal axis system of the EFG is defined by the local symmetry of the unit cell. For single crystalline samples a certain orientation of the crystal (and therewith of the EFG) versus the magnetic field can be chosen and well resolved, single resonance lines can be obtained and assigned as central lines or satellites (see for instance Fig. 2.2). By rotating the single crystal with respect to the applied magnetic field and observing the position changes of the satellites and the central line, it is possible to determine the orientation of the principal axes of the EFG and the parameters $\eta$ and $\nu_q \propto V_{ZZ}$. In polycrystalline samples however, each crystallite is randomly oriented to the external magnetic field. This leads to a distribution of Euler angles and thus to a broad powder pattern, whose shape can be calculated by summing up the contributions from all possible orientations in Equations (2.33) and (2.34). Additionally, possible anisotropies of the Knight shift tensor have to be considered.

For **the Knight shift tensor** defined in Eq. (2.14) there exists one coordinate system $(X', Y', Z')$ in which this tensor is diagonal. Let $\theta$ and $\phi$ be the angles describing the relative orientation of $(X', Y', Z')$ versus the direction of the external magnetic field along $z$, then the Knight shift measured along the direction $z$ of the applied magnetic field is given by [36, 55]:

$$K_z(\theta, \phi) = K_{X'} \sin^2\theta \cos^2\phi + K_{Y'} \sin^2\theta \sin^2\phi + K_{Z'} \cos^2\theta \tag{A.1}$$

To better express possible anisotropies, it is convenient to use [36, 52, 55]:

$$\begin{aligned} K_{iso} &= \frac{1}{3}(K_{X'} + K_{Y'} + K_{Z'}) \\ K_{aniso} &= \frac{1}{2}(K_{Y'} - K_{X'}) \\ K_{axial} &= \frac{1}{6}(2K_{Z'} - K_{X'} - K_{Y'}) \, , \end{aligned} \tag{A.2}$$

such that in the end the Knight shift $K_z$ along the direction $z$ of the applied magnetic field can be expressed as [36, 52, 55]:

$$K_z(\theta, \phi) = K_{iso} + K_{axial}(3\cos^2\theta - 1) + K_{aniso} \sin^2\theta \cos 2\phi \, . \tag{A.3}$$

The isotropic part $K_{iso}$ shifts the resonance line without affecting its line width, while the axial and the anisotropic contributions to the Knight shift, $K_{axial}$ and $K_{aniso}$, entail a broadening of the line.

**The first order quadrupole interactions** [see Eq. (2.33)] lead to the appearance of broadened quadrupolar satellites symmetrically distributed around the central transition [32, 36] (see Fig. A.1).

**In second order, the quadrupole interaction** [see Eq. (2.34)]:

$$\Delta^{(2)}\nu_m(\theta,\phi,\eta) = -\frac{\nu_q^2}{\nu_L}\frac{1}{6}\left[I(I+1)-\frac{3}{4}\right]\left[A(\phi,\eta)\cos^4\theta + B(\phi,\eta)\cos^2\theta + C(\phi,\eta)\right] \quad \text{(A.4)}$$

with the prefactors [54–56]:

$$\begin{aligned} A(\phi,\eta) &= -\frac{27}{8} - \frac{9}{4}\eta\cos 2\phi - \frac{3}{8}\eta^2\cos^2 2\phi \\ B(\phi,\eta) &= \frac{30}{8} - \frac{1}{2}\eta^2 + 2\eta\cos 2\phi + \frac{3}{4}\eta^2\cos^2 2\phi \\ C(\phi,\eta) &= -\frac{3}{8} + \frac{1}{3}\eta^2 + \frac{1}{4}\eta\cos 2\phi - \frac{3}{8}\eta^2\cos^2 2\phi \end{aligned} \quad \text{(A.5)}$$

gives rise to an asymmetric shape of the central line, where, depending on the value of $\eta$, five ($\eta < 1/3$) or six ($\eta > 1/3$) characteristic features can be observed [52, 55]. Fig. A.2 shows the line shape of the central transition for the case $\eta < 1/3$. Two singularities, two shoulders and a step appear as characteristic features in the line shape. For the prominent case of $\eta = 0$, the shoulders merge with their neighbouring singularities and the step appears at the Larmor frequency $\nu = \nu_L$ (neglecting magnetic hyperfine corrections, which will shift $\nu_L$ by the Knight shift $K$). If the second order quadrupole interaction is sufficiently large, it may also affect the satellites [52, 55]. For details of this effect, the reader is referred to [52, 53].

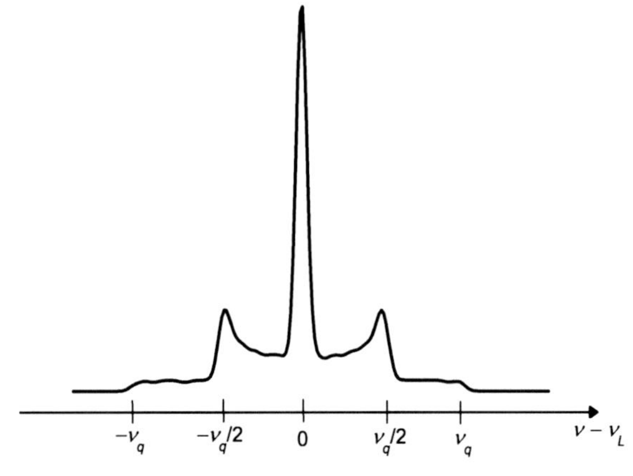


**Figure A.1:** Theoretical powder pattern for $I = 3/2$ in case of first order quadrupole effects for $\eta = 0$. For clarity the intensity of the satellites is artificially enhanced. Magnetic hyperfine contributions are neglected in this sketch ($K = 0$).

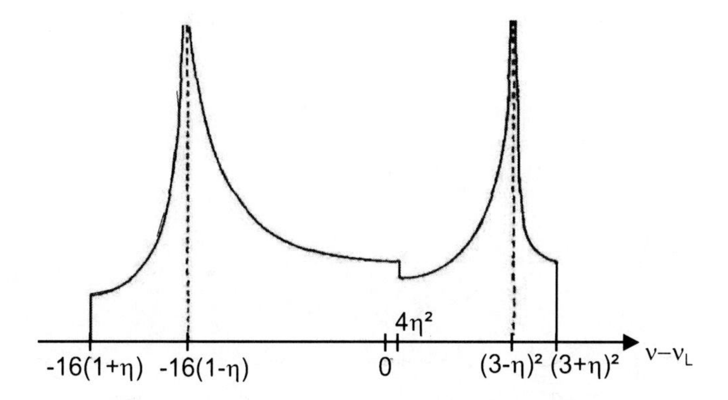


**Figure A.2:** Theoretical powder pattern for the central transition ($m = -{}^1\!/\!_2 \leftrightarrow m = +{}^1\!/\!_2$) in case of second order quadrupole effects for $\eta < 1/3$ (reproduced from [55]). Frequencies are given in units of $(\nu_q^2/144\nu_L)[I(I+1) - 3/4]$. Magnetic hyperfine interactions are neglected ($K = 0$).

**Combining second order quadrupole and first order magnetic hyperfine corrections** corresponding to Equations (A.3), (A.4) and (A.5), and assuming that the principal axes of the EFG and the Knight shift tensor coincide with each other, the powder pattern of the central resonance line resembles the one depicted in Fig. A.2. The characteristic features corresponding to the second order shifts as marked in Fig. A.2, are now additionally shifted against each other by the influence of the magnetic shift anisotropy. In general, the exact line shape of the powder pattern of the central line has to be computed including the mentioned effects. Since in powders the angles $\theta$ and $\phi$ are distributed with equal probability, the powder pattern can be calculated as the probability distribution of a resonance frequency $\nu$ in the interval $\nu + \mathrm{d}\nu$ within the solid angle $\mathrm{d}\cos\theta \mathrm{d}\phi$ [54, 55]. To extract the important quadrupole and magnetic shift parameters, it is sufficient to know the positions of the characteristic singularities and shoulders. These positions can be calculated by finding the critical points which solve:

$$\left(\frac{\partial \nu}{\partial \cos\theta}\right)_{\cos\theta = a, \phi = b} = \left(\frac{\partial \nu}{\partial \phi}\right)_{\cos\theta = a, \phi = b} = 0 \qquad \text{(A.6)}$$

on the surface $\nu = \nu(\cos\theta, \phi)$, where $(a, b)$ are the coordinates of a critical point [54, 55]. The surface $\nu = \nu(\cos\theta, \phi)$ is given by combining Equations (A.3), (A.4) and (A.5):

$$\nu(\theta, \phi) = [1 + K_z(\theta, \phi)]\nu_L + \Delta^{(2)}\nu(\theta, \phi)\,. \qquad \text{(A.7)}$$

An example of an experimentally observed powder pattern for the central transition, including second order quadrupole and first order magnetic hyperfine corrections is given in Fig. A.3. The black solid line is a simulation. It gives $\eta = 0$ and $\nu_q = \nu_c$ ($V_{ZZ} \parallel c$). Since $\eta = 0$, only two singularities and a step are visible as characteristic features in the spectrum. The shoulders as sketched in Fig. A.2 are merged with the singularities.

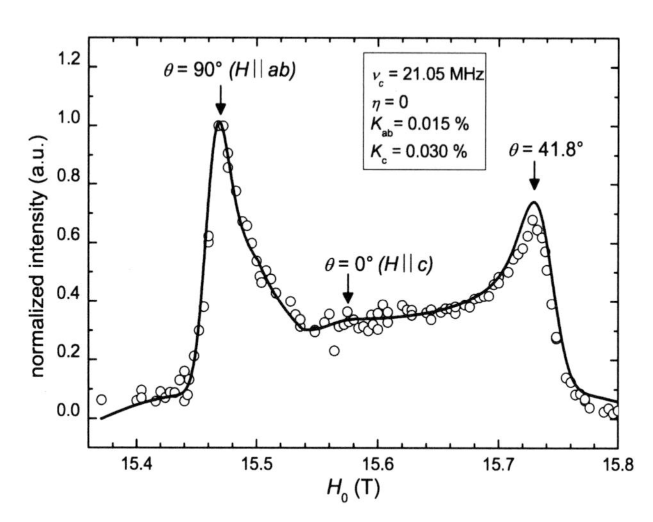


**Figure A.3:** Experimental powder pattern for the central transition of $^{75}$As NMR measured on a polycrystalline LiFeAs sample at $T = 100\,\mathrm{K}$. The black squares are the data points, the black solid line is a simulation of the powder spectrum including second order quadrupole effects and first order anisotropic magnetic hyperfine corrections. Values used for the simulation are given in the inset. The spectrum was measured by sweeping the magnetic field while applying a constant radio frequency $\nu$ and is thus mirror-inverted in comparison to Fig. A.2. Arrows denote the characteristic features of the powder pattern and the field orientation of the corresponding crystallites.

The value of $\nu_c = \nu_q = 21.05\,\mathrm{MHz}$ used for the simulation is close to $\nu_q = 21.18\,\mathrm{MHz}$ which was measured with $^{75}$As NQR (see fig. 8.2). The value of the spin shift along $ab$, $K_{ab} = 0.015\,\%$, is one order of magnitude lower than what has been measured in $^{75}$As NMR for $H \parallel ab$ (see Fig. 8.6). This might be due to the extreme sensitivity of $K$ versus the orientation of the external magnetic field (see Fig. 8.5 and 8.6).

## A.2 Calculation of Spin-Lattice Relaxation Functions

In the following, the general derivation of the relaxation function $M_z(t)$ will be reflected based on primary literature and overview articles [68–72, 309].

It starts with a consideration of the population of the nuclear magnetic energy levels $E_m$ and possible transitions between these levels. The time dependence of the normalized population $n_m$ of a nuclear magnetic energy level $E_m$ is expressed by [70]:

$$\frac{dn_m}{dt} = \sum_{n=1, n\neq m}^{N} (W_{mn}n_n - W_{nm}n_m) \qquad m = 1, 2, \ldots, N = 2I + 1 \,. \tag{A.8}$$

Written in a matrix notation, where $n_m$ is the $m$th component of the vector $\vec{n}$, this becomes:

$$\frac{d\vec{n}}{dt} = \mathbf{W}'\vec{n} \,, \tag{A.9}$$

where the components of the transition probability matrix $\mathbf{W}'$ are given by [70]:

$$W'_{mn} = \begin{cases} W_{mn} & \text{for } m \neq n \\ -\sum_{k,k\neq m} W_{km} & \text{for } m = n \,. \end{cases} \tag{A.10}$$

Due to the normalization $\sum_{m=1}^{N} n_m = 1$, the normalized populations are not independent from each other. It is therefore useful to define the population difference between adjacent energy levels as a new set of independent variables $\Delta n_m$, which are the $2I$ components of the vector $\overrightarrow{\Delta n}$:

$$\Delta n_m = n_m - n_{m+1} \quad m = 1, 2, \ldots, N - 1 = 2I \,. \tag{A.11}$$

The population difference $\Delta n_m$ is directly proportional to the NMR signal from the corresponding transition $E_{m+1} \leftrightarrow E_m$. In thermal equilibrium it will be defined as $\Delta n_m(t = \infty) =: \Delta n_{m,\infty}$, where $\Delta n_{m,\infty}$ is the $m$th component of the corresponding vector $\overrightarrow{\Delta n_\infty}$. Interesting for the theoretical description of the relaxation are the differences from the equilibrium values: $\Delta n_m(t) - \Delta n_{m,\infty}$. With these new variables, Eq. (A.9) becomes:

$$\frac{d(\overrightarrow{\Delta n}(t) - \overrightarrow{\Delta n_\infty})}{dt} = -\mathbf{A}\left(\overrightarrow{\Delta n}(t) - \overrightarrow{\Delta n_\infty}\right) , \tag{A.12}$$

where the components of the new matrix $\mathbf{A}$ are defined by:

$$-A_{mn} = (\mathbf{D}\mathbf{W}'\mathbf{D}^{-1})_{mn} \qquad m = 1, 2, \ldots, N - 1 = 2I \,. \tag{A.13}$$

The introduction of the transformation matrix $\mathbf{D}$ whose components read:

$$D_{mn} = \begin{cases} \delta_{mn} - \delta_{m+1,n} & \text{for } m \leq N - 1 \\ 1 & \text{for } m = N \,, \end{cases} \tag{A.14}$$

accounts for the transformation from Eq. (A.9) to Eq. (A.12) [70]. The solution of Eq. (A.12) is given by [70]:

$$\Delta n_m(t) = \Delta n_{m,\infty} + \sum_{n,k} C_{m,n} \exp(-\lambda_n t) C^{-1}_{n,k} [\Delta n_k(t = 0) - \Delta n_{k,\infty}] \,, \tag{A.15}$$

where $\lambda_n$ and $\mathbf{C}$ are the eigenvalues and the eigenvector matrix of the matrix $\mathbf{A}$, with the $n$th column of $\mathbf{C}$ being the eigenvector to the corresponding eigenvalue $\lambda_n$. The element $[\Delta n_k(t=0) - \Delta n_{k,\infty}]$ depends on the specific initial conditions.

Once the $2I$ solutions $\Delta n_m(t)$ are known, the relaxation function is expressed by [71]:

$$M(t) = M_0 \left[ 1 - \sum_m a_m \exp(-\lambda_m t) \right] , \tag{A.16}$$

where the prefactors $a_m$ depend on the eigenvector matrix $\mathbf{C}$ and the specific initial conditions expressed in $[\overrightarrow{\Delta n}(t=0) - \overrightarrow{\Delta n_\infty}]$. For an ideal inversion recovery measurement they have to fulfil the condition $\sum_m a_m = 2$. In the general case of a flip of the nuclear magnetization by an angle $\theta$, this condition becomes $\sum_m a_m = 1 - \cos\theta$. It is convenient to extract the inversion factor $f = \sum_m a_m$ out of the sum and keep normalized prefactors $a'_m$ with $\sum_m a'_m = 1$ in the sum. The general relaxation function then becomes:

$$\boxed{M(t) = M_0 \left[ 1 - f \left( \sum_m a'_m \exp(-\lambda_m t) \right) \right]}. \tag{A.17}$$

**The probabilities $W_{mn}$** for a transition between the eigenstates $|m\rangle$ and $|n\rangle$ of the static Hamiltonian $\mathcal{H}_0$ are in second-order perturbation theory given by [71]:

$$\begin{aligned} W_{mn} &\overset{m\neq n}{=} \frac{1}{\hbar^2} \int_{-\infty}^{\infty} dt \exp(i\omega_{mn} t) \overline{\langle m|\mathcal{H}_1(t)|n\rangle \langle n|\mathcal{H}_1(0)|m\rangle} \\ W_{mm} &= - \sum_{m\neq n} W_{mn} , \end{aligned} \tag{A.18}$$

where $\mathcal{H}_1$ is the perturbing Hamiltonian and $\omega_{mn} = (\langle m|\mathcal{H}_0|m\rangle - \langle n|\mathcal{H}_0|n\rangle)/\hbar$ is the transition frequency. The transition probability for a downward transition $W_{m-1\to m}$ is related to its corresponding upward transition $W_{m\to m-1}$ via the Boltzmann factor $\exp(\hbar\omega_0/k_B T)$. The exponent $\Delta = (\hbar\omega_0/k_B T)$ is usually very small, such that a series expansion can be used [68]: $W_{m-1\to m} = W_{m\to m-1}(1+\Delta)$. In the high temperature limit this gives $W_{m-1\to m} \approx W_{m\to m-1}$. Accordingly

$$W_{mn} \approx W_{nm} \tag{A.19}$$

is a good approximation for the considered temperature and frequency range[1].

Relaxation mechanisms may be induced by fluctuating magnetic fields $\vec{h}(t)$ and fluctuating components of the EFG $V_k(t)$. They are comprised in **the time-dependent perturbing Hamiltonian $\mathcal{H}_1(t)$** [71]:

$$\begin{aligned} \mathcal{H}_1(t) &= \mathcal{H}_{mag}(t) + \mathcal{H}_Q(t) \\ &\quad - \hbar\gamma\vec{I}\cdot\vec{h}(t) + \frac{eQ}{4I(2I-1)} \sum_{k=-2}^{2} V_k(t) T_{2k}(\vec{I}) , \end{aligned} \tag{A.20}$$

[1] Typical NMR frequencies lie in the range of tens to some hundreds of MHz, which correspond to temperatures in the mK range.

where $T_{2k}(\vec{I})$ are spherical tensor operators. Rapid nuclear spin exchange terms, which can also cause longitudinal relaxation, are omitted in $\mathcal{H}_1(t)$, since they proceed with times of the order of $T_2$ much smaller than $T_1$ and can therefore be neglected.

The static Hamiltonian $\mathcal{H}_0$ contains both magnetic and quadrupole contributions: $\mathcal{H}_0 = \mathcal{H}_Z + \mathcal{H}_Q$. As long as $\mathcal{H}_Z \gg \mathcal{H}_Q$, which is the case for $h\nu_{zz}\eta \ll \hbar\gamma H_0$,[2] the eigenfunctions of $\mathcal{H}_0$ can be approximated by the eigenfunctions of the Zeeman term $\mathcal{H}_Z$. The main magnetic and quadrupolar relaxation matrix terms are then given by [71]:

$$W_{mn}^{\text{mag}} = J(\omega_{mn})\{|\langle m|I^+|n\rangle|^2 + |\langle m|I^-|n\rangle|^2\} \tag{A.21}$$

$$W_{mn}^{\text{quad},1} = J^{(1)}(\omega_{mn})\{|\langle m|I^+I_z + I_zI^+|n\rangle|^2 + |\langle m|I^-I_z + I_zI^-|n\rangle|^2\} \tag{A.22}$$

$$W_{mn}^{\text{quad},2} = J^{(2)}(\omega_{mn})\{|\langle m|(I^+)^2|n\rangle|^2 + |\langle m|(I^-)^2|n\rangle|^2\} \tag{A.23}$$

with the spectral densities of the fluctuating magnetic and quadrupolar fields $J(\omega)$ and $J^{(1,2)}(\omega)$, respectively.

In the following, only the case of **pure magnetic relaxation** will be regarded. The spectral density for this case is given by:

$$J(\omega) = \frac{\gamma^2}{2}\int_{-\infty}^{\infty} dt \exp(i\omega_{mn}t)[h_+, h_-] \simeq J(0) =: W\,, \tag{A.24}$$

where $h_\pm = h_x \pm ih_y$ and $[A, B] = \frac{1}{2}\overline{(A(t)B(0) + B(t)A(0))}$. The last relation in Eq. (A.24) implies the supplementary assumption that the spectral density is well described by a single value[3]. From Eq. (A.21) it can be deduced, that in the case of magnetic relaxation only transitions between neighbouring energy levels $(m) \leftrightarrow (m-1)$ with $\Delta m = \pm 1$ are possible. The transition probability between two adjacent energy levels is then given by [68]:

$$W_{m\leftrightarrow m-1}^{\text{mag}} = W(I+m)(I-m+1)\,. \tag{A.25}$$

It has been shown [69] that for a magnetic relaxation of a given nuclear spin value $I$, there exist $2I$ relaxation times $\lambda_m^{-1}$ which amount to:

$$\lambda_m^{-1} = [p_m(p_m+1)W]^{-1} \quad \text{with } p_m = 1, 2, \ldots, 2I\,. \tag{A.26}$$

Which relaxation rates out of this series contribute to a certain relaxation functions depends on the specific transition $(m) \leftrightarrow (m-1)$ on which the relaxation is observed.

**For a nuclear spin $I = 3/2$** the transition probabilities are $W_{-1/2\leftrightarrow-3/2} = W_{+3/2\leftrightarrow+1/2} = 3W$ and $W_{+1/2\leftrightarrow-1/2} = 4W$. The $2I$ relaxation times amount to $\lambda_{-1/2} = 2W$, $\lambda_{1/2} = 6W$ and $\lambda_{3/2} = 12W$. Supposing that the system was in thermal equilibrium before being exposed to an inversion recovery pulse sequence, where the 180° pulse was short enough for the relaxation during the pulse to be negligible, the relaxtion of the central resonance line $(m = +1/2) \leftrightarrow (m = -1/2)$ is described by:

$$M_z(t) = M_0\{1 - f[0.9\exp(-12Wt) + 0.1\exp(-2Wt)]\}\,, \tag{A.27}$$

[2] The condition $h\nu_{zz}\eta = \hbar\omega_{zz}\eta \ll \hbar\gamma H_0$ is fulfilled for small quadrupole frequencies $\nu_{zz}$ as well as for large quadrupole frequencies $\nu_{zz}$ as long as the external field $H_0$ is aligned along the direction of the principal axis of the EFG ($\eta \ll 1$) [70]. If the measurement conditions do not comply with the mentioned restriction, the analysis is more complicated.

[3] The spectral density can be described by a single value $W$ as long as the inverse correlation time $\tau_c^{-1}$ of the fluctuating magnetic field $h(t)$ is large compared to $\omega_{mn}$: $\omega_{mn}\tau_c \ll 1$.

while on the satellite transitions $(m = \pm 3/2) \leftrightarrow (m = \pm 1/2)$ the recovery of the magnetization follows:

$$M_z(t) = M_0\{1 - f[0.4\exp(-12Wt) + 0.5\exp(-6Wt) + 0.1\exp(-2Wt)]\}\,. \qquad \text{(A.28)}$$

**The spin-lattice relaxation time $T_1$** is arbitrarily defined. It is common to use:

$$\frac{1}{T_1} = 2W\,, \qquad \text{(A.29)}$$

but also other definitions of $T_1$ relative to $W$ or some of the $W_{m \leftrightarrow m-1}$ are possible. Using the definition (A.29) and the equations (A.27) and (A.28) one obtains the relaxation functions (2.54) and (2.55), which were used in the course of this work for NMR measurements.

If there exist also relevant quadrupole contributions to the relaxation, additional quadrupole transition probabilities $W_{m \leftrightarrow n}^{\text{quad},(1),(2)}$ will have to be considered, which are multiples of the spectral densities of the fluctuating components of the EFG defined by [71]:

$$J^{(1,2)}(\omega) = \frac{(eQ)^2}{\hbar^2} \int_{-\infty}^{\infty} dt \exp(i\omega_{mn}t)[V_{+1,2}, V_{-1,2}] \simeq J^{(1,2)}(0) =: W_{1,2}\,, \qquad \text{(A.30)}$$

where $V_{\pm 1} = V_{xz} \pm iV_{yz}$ and $V_{\pm 2} = \frac{1}{2}(V_{xx} - V_{yy}) \pm iV_{xy}$ [310]. $W_1$ and $W_2$ refer to quadrupole transitions with $\Delta m = \pm 1$ and $\Delta m = \pm 2$, respectively. In this case, one has three defined spin-lattice relaxation times $T_1$, $T_1^{(1)}$ and $T_1^{(2)}$ related to $W$, $W_1$ and $W_2$. The relaxation cannot be described by one specifically defined spin-lattice relaxation time $T_1$ any more.

## A.3 Stretched Exponential Relaxation Function

Stretching exponents $\lambda < 1$ represent a phenomenological expression for the distribution of spin-lattice relaxation rates around a characteristic time $T_1$. In Section 7.2 we used a stretched exponential function to describe the inversion recovery curves of $LaO_{0.9}F_{0.1}FeAs_{1-\delta}$ in the temperature range $T_c \geqslant T > 14\,\mathrm{K}$:

$$M_z(t) = M_0[1 - f(0.9\mathrm{e}^{-(6\mathrm{t}/\mathrm{T}_1)^\lambda} + 0.1\mathrm{e}^{-(\mathrm{t}/\mathrm{T}_1)^\lambda})]\,. \tag{A.31}$$

However, there exists no generally accepted practice of the use of the stretching exponent $\lambda$ in multi-exponential relaxation functions of nuclear spins $I > 1/2$. In the following, the use of a stretching exponent $\lambda$ in the literature will therefore be recapitulated and the choice of Eq. (A.31) will be explained.

For recoveries of the nuclear magnetization of a nuclear spin of $I = 1/2$ or a nuclear spin of $I = 3/2$ with a non-existing or negligibly small quadrupole splitting, such that all three resonances are excited simultaneously, the use of a stretching exponent $\lambda$ is quite common and straightforward [311–313]. The relaxation function reads:

$$M_z(t) = M_0[1 - f\mathrm{e}^{-(\mathrm{t}/\mathrm{T}_1)^\lambda}] \tag{A.32}$$

and does not cause difficulties in the quantitative comparison and interpretation, since the normal relaxation function is given by exactly the same equation with $\lambda = 1$.

For multi-exponential relaxation functions, which are needed for nuclear spins $I > 1/2$, three trends could be found in literature:

**1.** Some references, all dealing with a nuclear spin of $I \geqslant 7/2$, refer to the very simple stretched exponential function of the form of Eq. (A.32) [66, 314–317]. Most of them fixed the stretching exponent to $\lambda = 0.5$. The choice of Eq. (A.32) is arbitrary since the standard relaxation function of a nuclear spin of $I = 7/2$ is normally multi-exponential with a single $T_1$ [318]:

$$M_z(t) = M_0[1 - (A\mathrm{e}^{-28\mathrm{t}/\mathrm{T}_1} + \mathrm{B}\mathrm{e}^{-15\mathrm{t}/\mathrm{T}_1} + \mathrm{C}\mathrm{e}^{-6\mathrm{t}/\mathrm{T}_1} + \mathrm{D}\mathrm{e}^{-\mathrm{t}/\mathrm{T}_1})]\,. \tag{A.33}$$

The switching between the use of the multi-exponential Eq. (A.33) and the stretched single exponential formula Eq. (A.32) in certain temperature regions leads to unphysical steps in the temperature dependence of $T_1^{-1}$, since Eq. (A.32) neglects the multi-exponential nature of the relaxation. Thus the values of $T_1$ derived from Eq. (A.32) are much too short. The overall qualitative temperature dependence will however be the same.

**2.** Other authors tried to include the stretching exponent $\lambda$ directly into the multi-exponential relaxation functions for $I = 5/2$ and $I = 7/2$, respectively [319, 320], leading to multiple exponents of the form (const. $\times\ t/T_1)^\lambda$, similar to the relaxation function Eq. (A.31) which was used in Chapter 7.2.[4] The denotation of $\lambda$ is the same as in

[4] Note that some authors of [320] presented seven years earlier a relaxation function with exponents of the form [const. $\times\ (t/T_1)^\lambda$], where the numerical coefficients const. are not raised to the power of $\lambda$ [318]. This is however unphysical, since the choice of $T_1$ is rather arbitrary (see discussion of the calculation of the relaxation functions in Appendix A.2). Depending on the theoretical definitions, it might be $T_1 = 3/(2W_1)$ or $T_1 = 1/(2W_1)$, where $W_1$ is a specific magnetic relaxation rate of the nuclear spin system, which also has to be defined [72]. Therefore the numerical coefficients have to be included in the radix. The frequent use of the expression $[(3t/T_1)^\lambda]$ in the relaxation functions of NQR experiments on $I = 3/2$ [29, 30, 49, 321–325] is another example of the use of stretching exponent $\lambda$ including numerical prefactors.

the former case. The advantage of this method is that the absolute values of $T_1^{-1}$ are comparable between fits with $\lambda = 1$ and with $\lambda < 1$ and thus no artificial steps will appear in the temperature dependence of $T_1^{-1}$, when switching from $\lambda = 1$ to $\lambda < 1$.

**3.** A distribution of spin-lattice relaxation rates can even be modelled, leading to relaxation functions of the type [72, 311, 320, 326]:

$$M_z(t) = \int P(\omega) M_{z,\lambda=1}(\omega, t) d\omega \,, \tag{A.34}$$

where $P(\omega)$ is a chosen probability distribution function (Gaussian, rectangular, Lorentzian,..) of nuclei relaxing at a rate $\omega$ and $M_{z,\lambda=1}(\omega, t)$ is the corresponding multi-exponential relaxation function. $P(\omega)$ has to be normalized to unity:

$$\int_0^\infty P(\omega) d\omega = 1 \,. \tag{A.35}$$

Using for instance a Gaussian distribution, the fitting function Eq. (A.34) is determined by only two parameters: the center of the Gaussian distribution, which displays the most probable relaxation rate $T_1^{-1}$, and the width $\sigma$ of the Gaussian distribution. It has been shown that the width of the Gaussian distribution $\sigma$ is directly proportional to the stretching exponent $\lambda$ of a stretched multi-exponential relaxation function $M_z[(\text{const.} \times t/T_1)^\lambda]$ as described in case **2.** and that the absolute $T_1^{-1}$ values derived from both fitting procedures are very similar [320].

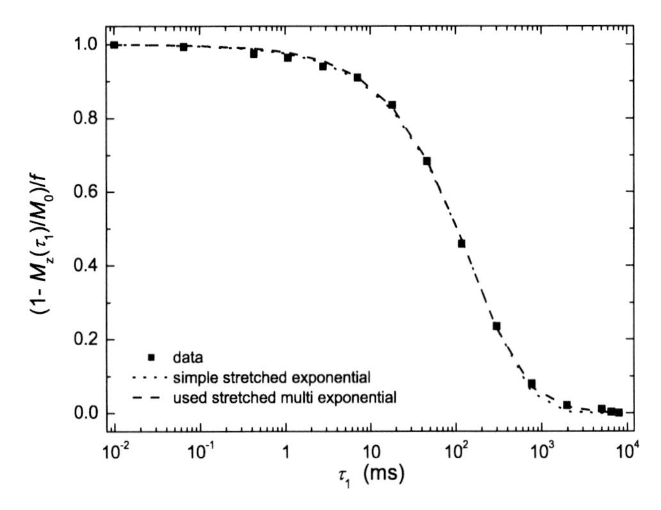


**Figure A.4:** Stretched exponential fitting for $LaO_{0.9}F_{0.1}FeAs_{1-\delta}$ at $T = 22\,K$ in $H = 7.01\,T$. Black squares denote the data points. The dotted/dashed line represents fitting with Eq. (A.32)/Eq. (A.31), respectively.

| $T$ (K) | $T_1$ (ms) from (A.32) | $T_1$ (ms) from (A.31) | ratio (A.31) /(A.32) |
|---|---|---|---|
| 24 | 116.8 | 595.2 | 5.1 |
| 22 | 174.9 | 896.3 | 5.1 |
| 20 | 307.5 | 1617.9 | 5.3 |
| 18 | 530.4 | 2787.9 | 5.3 |
| 16 | 764.0 | 3934.2 | 5.1 |

**Table A.1:** $T_1$ values for $LaO_{0.9}F_{0.1}FeAs_{1-\delta}$ derived from the different stretched exponential fitting functions (A.32) and (A.31) in the temperature range $T_c \geqslant T > 14\,\mathrm{K}$ and their corresponding ratio.

Just for comparison, the recovery curves were also fit with the simple single stretched exponential function (A.32). Both fitting procedures (A.31) and (A.32) gave satisfactory fits with coefficients of determination ($R^2$) higher than 0.999. Figure A.4 shows the two fitting functions for the data at $T = 22\,\mathrm{K}$ in $H = 7.01\,\mathrm{T}$. The fits are nearly not distinguishable by eye. They yielded however very different $T_1$ values, which are summarized in table A.1 for the whole relevant temperature range.

For the fitting of the recovery curves of $LaO_{0.9}F_{0.1}FeAs_{1-\delta}$ in the very small temperature range $T_c \geqslant T > 14\,\mathrm{K}$ Eq. (A.31) corresponding to approach **2.** was preferred to the other two possibilities, since Eq. (A.32) (case **1.**) would have yielded artificially small $T_1$ values and an approach as in case **3.** would not have been of additional benefit, since it was the $T_1$ values which were from interest and not their concrete distribution function $P(\omega)$.

The values differ in a fixed factor of 5.2 from each other. The resulting temperature dependence of $T_1^{-1}$ is therefore unaffected by the choice of the fitting function. Only the absolute values changed. This is summarized in Fig. A.5. The discussion in this Chapter shows that one always have to be careful when comparing $T_1$ values quantitatively with the results of other groups or measurements, since the absolute values depend on the underlying definition of the relaxation rate $W_1$ and its relation to $T_1$ [72].

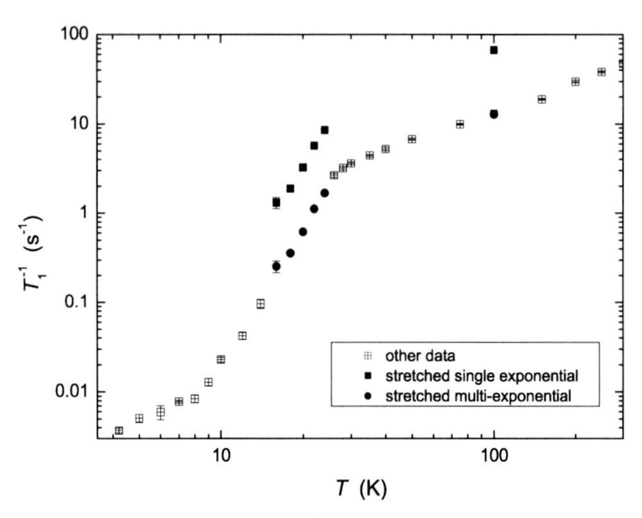


**Figure A.5:** Spin-lattice relaxation rate $T_1^{-1}$ of $LaO_{0.9}F_{0.1}FeAs_{1-\delta}$ in $H = 7.01\,T$ for different stretched exponential fitting functions in the temperature range $T_c \geqslant T > 14\,K$ and (for comparison) at $T = 100\,K$ with $\lambda \approx 1$. Open squares denote the data derived from fitting with (7.1) [(7.2)] for $T > T_c$ [$T \leq 14\,K$]. Filled squares are data fitted from Eq. (A.32) and dots from Eq. (A.31). The qualitative temperature dependence is the same while the absolute values are quite different (note the double logarithmic scale).

## A.4 Spin Diffusion

Among various possible mechanisms for the linear temperature dependence of $T_1^{-1}$ at very low temperature ($T < 0.3T_c$) in the superconducting state of $LaO_{0.9}F_{0.1}FeAs_{1-\delta}$, the classical spin diffusion from vortex cores was listed in Section 7.2. Spin diffusion itself is the diffusion of the nuclear magnetization among the nuclear spins due to their dipole-dipole coupling [33]. In superconductors in the flux-melted phase, spin diffusion may also occur via diffusion of the vortex cores [327]. If the spin-spin relaxation time $T_2$ is much shorter than the spin-lattice relaxation time $T_1$, the spin-lattice relaxation in the (normal conducting) vortex cores $T_{1,n}^{-1}$ and in the superconducting intervortex regions $T_{1,sc}^{-1}$ average together to the observed $T_{1,obs}^{-1}$ in the superconducting state. The effect of spin diffusion from vortex cores is only observable as long as $T_{1,n}^{-1} \gg T_{1,sc}^{-1}$ [328]. This condition is fullfilled at low temperature, since $T_{1,sc}^{-1}$ decreases fast in the superconducting state because of the opening of the superconducting energy gap. For this reason, the effect of spin diffusion on the spin-lattice relaxation rate can only be visible at low temperature. In this regime, the measured $T_{1,obs}^{-1}$ is higher than the intrinsic spin-lattice relaxation $T_{1,sc}^{-1}$ and therefore leads to a deviation from the intrinsic temperature dependence of $T_{1,sc}^{-1}$. A recent example is the conventional $s_{++}$-multiband superconductor $YNi_2B_2C$ [292]. Spin diffusion may also arise in more or less clean $s_{\pm}$-superconductors, but not in unconventional superconductors with pronounced nodes or many impurity-induced states in the gap.

Although the lack of field-dependence of $(T_1T)^{-1}$ of $LaO_{0.9}F_{0.1}FeAs_{1-\delta}$ revealed that spin diffusion can not be the reason for $T_1^{-1} \propto T$ at low temperature, in the following the analysis of the first NMR data in 7.01 T in terms of the spin diffusion mechanism will be presented, since it was a prominent candidate awhile.

In the case of spin diffusion, the ratio of of $(T_1T)^{-1}$ in the normal state and below $0.3\,T_c$ has to be compared with the upper critical field $H_{c2}^*$ scaled by the field $H$ of the measurements [328, 329]:

$$\alpha = \frac{(T_1T)_{obs}^{-1}}{(T_1T)_n^{-1}} = \frac{H}{\Phi_0}\xi_{ab}^2 = \frac{H}{2\pi H_{c2}^*} . \tag{A.36}$$

In this equation $\Phi_0$ denotes the flux quantum and $\xi_{ab}$ the coherence length. $\frac{H}{\Phi_0}\xi_{ab}^2$ is the volume fraction of the vortex cores. Eq. (A.36) is only valid under the condition $H_{c1} \ll H \ll H_{c2}$, which is fullfilled for the discussed measurements.

Fig. A.6 shows the spin-lattice relaxation rate $T_1^{-1}$ of $LaO_{0.9}F_{0.1}FeAs_{1-\delta}$ versus temperature. The blue line is a fit to a linear temperature dependence at high temperature. The inset shows the data below 0.3 $T_c$ and the corresponding linear fit (red line). The comparison of both slopes yielded $\alpha = \frac{(T_1T)_{obs}^{-1}}{(T_1T)_n^{-1}} \approx 0.008$. This gives a coherence length of $\xi_{ab} \approx 15.4\,\text{Å}$, which corresponds to an orbital upper critical field of $H_{c2}^*(0) = 140\,\text{T}$, in reasonable agreement with $H_{c2}^*(0) = (119 \pm 13)\,\text{T}$ derived from the measured slopes $\mathrm{d}H_{c2}/\mathrm{d}T\,|_{T_c}$ [219, 290].

The shaded yellow area in Fig. A.6 is the estimate of the slope of $T_1^{-1}$ at high temperatures, by taking the fitted slope of $T_1^{-1}$ at low temperatures and $H_{c2}^*(0) = (119 \pm 13)\,\text{T}$ as it was estimated from resistivity data [219, 290]. The boundaries of this area are given by the error of the upper critical field. Both ways of analyzing the data seem to

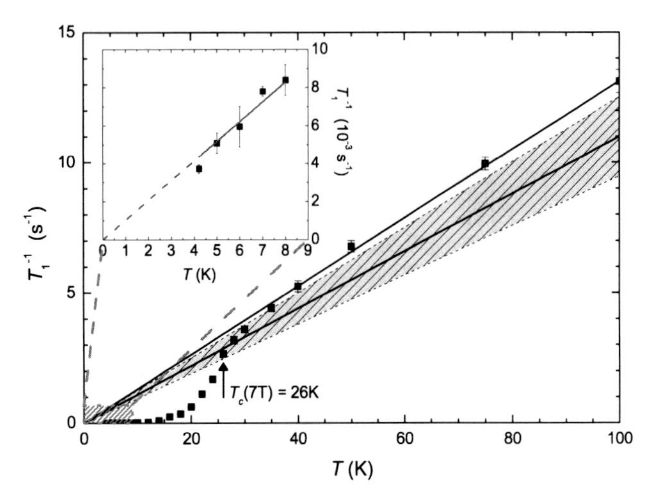


**Figure A.6:** $T_1^{-1}$ of $\mathrm{LaO_{0.9}F_{0.1}FeAs_{1-\delta}}$ versus temperature measured in $H_0 = 7.01\,\mathrm{T}$. The inset highlights the linear temperature dependence in the superconducting state. The blue (red) line is a linear fit to the data in the normal (superconducting) state. The yellow shaded area shows the expected slope of $(T_1T)_n^{-1}$ in the normal state for an upper critical field of $H_{c2}^*(0) = (119 \pm 13)\,\mathrm{T}$. The arrow denotes $T_c(H_0) = 26\,\mathrm{K}$.

work reasonably. But, as already stated above, subsequent measurements revealed that a field dependence of $(T_1T)^{-1}$ was missing. Fig. A.7 shows the field dependence of $(T_1T)_{obs}^{-1}$ at $T = 8\,\mathrm{K}$. Taking into consideration the field independence of $(T_1T)_n^{-1}$ (see fig. 7.7), $(T_1T)_{obs}^{-1}$ should scale linearly with $H$ [see Eq. (A.36)]. This is depicted by the black line in Fig. A.7, using the formerly obtained value of $H_{c2}^*(0) = 140\,\mathrm{T}$. The data do not follow the required linear field dependence. Spin diffusion can therefore be ruled out.

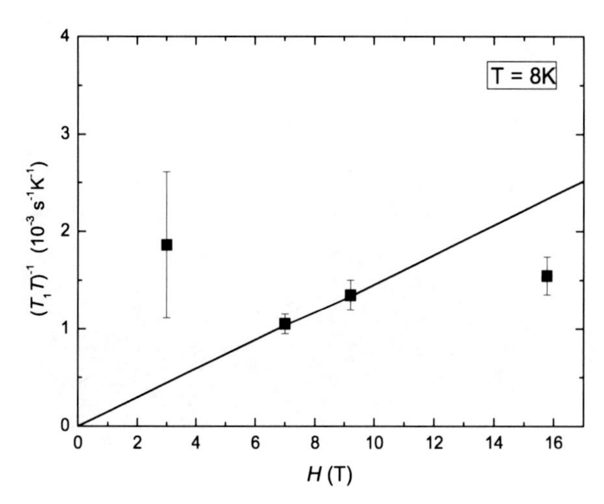


**Figure A.7:** Field dependence of $(T_1T)^{-1}_{sc}$ at $T = 8$ K of $LaO_{0.9}F_{0.1}FeAs_{1-\delta}$. The line is the expected field dependence in the case of spin diffusion which gives: $(T_1T)^{-1}_{sc} = \frac{(T_1T)^{-1}_n \xi^2_{ab}}{\Phi_0} H$.

# Bibliography


[1] H. Kamerlingh Onnes, *Commun. Phys. Lab. Univ. Leiden* **12**, 120 (1911).

[2] J. Bardeen, L. N. Cooper, and J. R. Schrieffer, *Phys. Rev.* **108**, 1175 (1957).

[3] F. Steglich, J. Aarts, C. D. Bredl, W. Lieke, D. Meschede, W. Franz, and H. Schäfer, *Phys. Rev. Lett.* **43**, 1892 (1979).

[4] J. G. Bednorz and K. A. Müller, *Zeitschrift für Physik B Condensed Matter* **64**, 189 (1986).

[5] Y. Kamihara, T. Watanabe, M. Hirano, and H. Hosono, *J. Am. Chem. Soc.* **130**, 3296 (2008).

[6] Z.-A. Ren, J. Yang, W. Lu, W. Yi, X.-L. Shen, Z.-C. Li, G.-C. Che, X.-L. Dong, L.-L. Sun, F. Zhou, and Z.-X. Zhao, *Europhys. Lett.* **82**, 57002 (2008).

[7] Z.-A. Ren, W. Lu, J. Yang, W. Yi, X.-L. Shen, Zheng-Cai, G.-C. Che, X.-L. Dong, L.-L. Sun, F. Zhou, and Z.-X. Zhao, *Chinese Physics Letters* **25**, 2215 (2008).

[8] Z.-A. Ren, G.-C. Che, X.-L. Dong, J. Yang, W. Lu, W. Yi, X.-L. Shen, Z.-C. Li, L.-L. Sun, F. Zhou, and Z.-X. Zhao, *Europhys. Lett.* **83**, 17002 (2008).

[9] X. H. Chen, T. Wu, G. Wu, R. H. Liu, H. Chen, and D. Fang, *Nature* **453**, 761 (2008).

[10] K. Ishida, Y. Nakai, and H. Hosono, *J. Phys. Soc. Jpn.* **78**, 062001 (2009).

[11] M. Rotter, M. Tegel, D. Johrendt, I. Schellenberg, W. Hermes, and R. Pöttgen, *Phys. Rev. B* **78**, 020503 (2008).

[12] M. Rotter, M. Tegel, and D. Johrendt, *Phys. Rev. Lett.* **101**, 107006 (2008).

[13] P. L. Alireza, Y. T. C. Ko, J. Gillett, C. M. Petrone, J. M. Cole, G. G. Lonzarich, and S. E. Sebastian, *Journal of Physics: Condensed Matter* **21**, 012208 (2009).

[14] J. H. Tapp, Z. Tang, B. Lv, K. Sasmal, B. Lorenz, P. C. W. Chu, and A. M. Guloy, *Phys. Rev. B* **78**, 060505 (2008).

[15] X. Wang, Q. Liu, Y. Lv, W. Gao, L. Yang, R. Yu, F. Li, and C. Jin, *Solid State Communications* **148**, 538 (2008).

[16] M. J. Pitcher, D. R. Parker, P. Adamson, S. J. C. Herkelrath, A. T. Boothroyd, R. M. Ibberson, M. Brunelli, and S. J. Clarke, *Chem. Commun.* , 5918 (2008).

[17] F.-C. Hsu, J.-Y. Luo, K.-W. Yeh, T.-K. Chen, T.-W. Huang, P. M. Wu, Y.-C. Lee, Y.-L. Huang, Y.-Y. Chu, D.-C. Yan, and M.-K. Wu, *Proceedings of the National Academy of Sciences* **105**, 14262 (2008).

[18] L. C. Hebel and C. P. Slichter, *Phys. Rev.* **107**, 901 (1957).

[19] L. C. Hebel and C. P. Slichter, *Phys. Rev.* **113**, 1504 (1959).

[20] Y. Masuda and A. G. Redfield, *Phys. Rev.* **125**, 159 (1962).

[21] K. Ishida, H. Mukuda, Y. Kitaoka, K. Asayama, Z. Q. Mao, Y. Mori, and Y. Maeno, *Nature* **396**, 658 (1998).

[22] H. Tou, Y. Kitaoka, K. Asayama, N. Kimura, Y. Ōnuki, E. Yamamoto, and K. Maezawa, *Phys. Rev. Lett.* **77**, 1374 (1996).

[23] C. Berthier, M. H. Julien, M. Horvatić, and Y. Berthier, *J. Phys. I France* **6**, 2205 (1996).

[24] E. M. Purcell, H. C. Torrey, and R. V. Pound, *Phys. Rev.* **69**, 37 (1946).

[25] F. Bloch, W. W. Hansen, and M. Packard, *Phys. Rev.* **69**, 127 (1946).

[26] C. Gorter and L. Broer, *Physica* **9**, 591 (1942).

[27] G. Schatz and A. Weidinger, *Nukleare Festkörperphysik*, B.G. Teubner, Stuttgart, 1992.

[28] C. P. Slichter, *Principles of Magnetic Resonance*, Springer, Berlin, Heidelberg, New York, 1990.

[29] H.-J. Grafe, F. Hammerath, A. Vyalikh, G. Urbanik, V. Kataev, T. Wolf, G. Khaliullin, and B. Büchner, *Phys. Rev. B* **77**, 014522 (2008).

[30] G. Lang, H.-J. Grafe, D. Paar, F. Hammerath, K. Manthey, G. Behr, J. Werner, and B. Büchner, *Phys. Rev. Lett.* **104**, 097001 (2010).

[31] M. Takigawa, A. P. Reyes, P. C. Hammel, J. D. Thompson, R. H. Heffner, Z. Fisk, and K. C. Ott, *Phys. Rev. B* **43**, 247 (1991).

[32] A. Abragam, *Principles of Nuclear Magnetism*, Oxford University Press, New York, 1961.

[33] E. Fukushima and S. B. W. Roeder, *Experimental Pulse NMR - A Nuts and Bolts Approach*, Westview Press, 1981.

[34] H.-J. Grafe, *Nuclear Magnetic Resonance Studies on Rare Earth Co-doped Lanthanum Cuprates*, PhD thesis, TU Dresden, 2005.

[35] A. U. B. WOLTER, *Longitudinal and Transverse Magnetization in Low-Dimensional Molecule-Based Quantum Magnets*, PhD thesis, TU Braunschweig, 2005.

[36] G. LANG, *Étude par RMN et μSR des propriétés électroniques et magnétiques des cobaltates de sodium*, PhD thesis, Université Paris XI Orsay, 2007.

[37] S.-H. BAEK, *Spin dynamics and quantum tunneling in Fe8 nanomagnet and in AFM rings by NMR*, PhD thesis, Iowa State University, 2004.

[38] N. J. CURRO, *Nuclear Magnetic Resonance Studies of the High Temperature Superconductors in the normal and superconducting states*, PhD thesis, University of Illinois at Urbana-Champaign, 1998.

[39] *www.webelements.com* .

[40] M. POGGIO and C. L. DEGEN, *Nanotechnology* **21**, 342001 (2010).

[41] I. BAKONYI, I. KOVÁCS, and I. PÓCSIK, *physica status solidi (b)* **114**, 609 (1982).

[42] N. F. RAMSEY and E. M. PURCELL, *Phys. Rev.* **85**, 143 (1952).

[43] N. F. RAMSEY, *Phys. Rev.* **91**, 303 (1953).

[44] M. A. RUDERMAN and C. KITTEL, *Phys. Rev.* **96**, 99 (1954).

[45] N. BLOEMBERGEN and T. J. ROWLAND, *Phys. Rev.* **97**, 1679 (1955).

[46] M. TAKIGAWA, *Phys. Rev. B* **49**, 4158 (1994).

[47] D. THELEN and D. PINES, *Phys. Rev. B* **49**, 3528 (1994).

[48] C. H. PENNINGTON and C. P. SLICHTER, *Phys. Rev. Lett.* **66**, 381 (1991).

[49] M. TAKIGAWA, N. MOTOYAMA, H. EISAKI, and S. UCHIDA, *Phys. Rev. B* **57**, 1124 (1998).

[50] A. J. MILLIS, H. MONIEN, and D. PINES, *Phys. Rev. B* **42**, 167 (1990).

[51] W. D. KNIGHT, *Phys. Rev.* **76**, 1259 (1949).

[52] R. B. CREEL, S. L. SEGEL, R. J. SCHOENBERGER, R. G. BARNES, and D. R. TORGESON, *The Journal of Chemical Physics* **60**, 2310 (1974).

[53] R. G. BARNES, D. J. GENIN, R. G. LECANDER, and D. R. TORGESON, *Phys. Rev.* **145**, 302 (1966).

[54] K. NARITA, J. UMEDA, and H. KUSUMOTO, *The Journal of Chemical Physics* **44**, 2719 (1966).

[55] J. F. BAUGHER, P. C. TAYLOR, T. OJA, and P. J. BRAY, *The Journal of Chemical Physics* **50**, 4914 (1969).

[56] D. FREUDE, J. HAASE, J. KLINOWSKI, T. A. CARPENTER, and G. RONIKIER, *Chemical Physics Letters* **119**, 365 (1985).

[57] G. M. VOLKOFF, *Canad. J. Phys.* **31**, 820 (1953).

[58] R. V. POUND, *Phys. Rev.* **79**, 685 (1950).

[59] E. C. REYNHARDT, *Journal of Physics C: Solid State Physics* **7**, 4135 (1974).

[60] W. G. PROCTOR and W. H. TANTTILA, *Phys. Rev.* **98**, 1854 (1955).

[61] F. BLOCH, *Phys. Rev.* **70**, 460 (1946).

[62] N. BLOEMBERGEN, E. M. PURCELL, and R. V. POUND, *Nature* **160**, 475 (1947).

[63] N. BLOEMBERGEN, E. M. PURCELL, and R. V. POUND, *Phys. Rev.* **73**, 679 (1948).

[64] K. H. KIM, D. R. TORGESON, F. BORSA, J. CHO, S. W. MARTIN, and I. SVARE, *Solid State Ionics* **91**, 7 (1996).

[65] B. J. SUH, P. C. HAMMEL, M. HÜCKER, B. BÜCHNER, U. AMMERAHL, and A. REVCOLEVSCHI, *Phys. Rev. B* **61**, R9265 (2000).

[66] N. J. CURRO, P. C. HAMMEL, B. J. SUH, M. HÜCKER, B. BÜCHNER, U. AMMERAHL, and A. REVCOLEVSCHI, *Phys. Rev. Lett.* **85**, 642 (2000).

[67] T. MORIYA, *J. Phys. Soc. Jpn.* **18**, 516 (1963).

[68] E. R. ANDREW and K. M. SWANSON, *Proceedings of the Physical Society* **75**, 582 (1960).

[69] E. R. ANDREW and D. P. TUNSTALL, *Proceedings of the Physical Society* **78**, 1 (1961).

[70] M. HORVATIĆ, *Journal of Physics: Condensed Matter* **4**, 5811 (1992).

[71] A. SUTER, M. MALI, J. ROOS, and D. BRINKMANN, *Journal of Physics: Condensed Matter* **10**, 5977 (1998).

[72] C. T. MILLING, *Nuclear Magnetic Resonance Studies of the Effect of Vortices on the High Temperature Superconductor $YBa_2Cu_3O_7$*, PhD thesis, University of Illinois at Urbana-Champaign, 2001.

[73] J. CHEPIN and J. H. ROSS JR., *Journal of Physics: Condensed Matter* **3**, 8103 (1991).

[74] D. E. MACLAUGHLIN, J. D. WILLIAMSON, and J. BUTTERWORTH, *Phys. Rev. B* **4**, 60 (1971).

[75] J. KORRINGA, *Physica* **16**, 601 (1950).

[76] A. NARATH and H. T. WEAVER, *Phys. Rev.* **175**, 373 (1968).

[77] B. S. Shastry and E. Abrahams, *Phys. Rev. Lett.* **72**, 1933 (1994).

[78] N. J. Curro, *Reports on Progress in Physics* **72**, 026502 (2009).

[79] H.-J. Grafe, D. Paar, G. Lang, N. J. Curro, G. Behr, J. Werner, J. Hamann-Borrero, C. Hess, N. Leps, R. Klingeler, and B. Büchner, *Phys. Rev. Lett.* **101**, 047003 (2008).

[80] P. Jeglič, A. Potočnik, M. Klanjšek, M. Bobnar, M. Jagodič, K. Koch, H. Rosner, S. Margadonna, B. Lv, A. M. Guloy, and D. Arčon, *Phys. Rev. B* **81**, 140511 (2010).

[81] Z. Li, Y. Ooe, X.-C. Wang, Q.-Q. Liu, C.-Q. Jin, M. Ichioka, and G. qing Zheng, *J. Phys. Soc. Jpn.* **79**, 083702 (2010).

[82] G. Lang, J. Bobroff, H. Alloul, G. Collin, and N. Blanchard, *Phys. Rev. B* **78**, 155116 (2008).

[83] A. G. Redfield, *Phys. Rev.* **98**, 1787 (1955).

[84] R. E. Walstedt, *Phys. Rev. Lett.* **19**, 146 (1967).

[85] R. E. Walstedt and S.-W. Cheong, *Phys. Rev. B* **51**, 3163 (1995).

[86] Z. Yamani, B. W. Statt, W. A. MacFarlane, R. Liang, D. A. Bonn, and W. N. Hardy, *Phys. Rev. B* **73**, 212506 (2006).

[87] D. E. MacLaughlin, *Magnetic Resonance in the Superconducting State* in *Solid State Physics* by H. Ehrenreich *et. al.* (Eds.), Vol. 31, Academic Press, New York, 1976.

[88] K. Yosida, *Phys. Rev.* **110**, 769 (1958).

[89] A. P. Mackenzie and Y. Maeno, *Rev. Mod. Phys.* **75**, 657 (2003).

[90] H. Murakawa, K. Ishida, K. Kitagawa, Z. Q. Mao, and Y. Maeno, *Phys. Rev. Lett.* **93**, 167004 (2004).

[91] F. Reif, *Phys. Rev.* **106**, 208 (1957).

[92] A. M. Clogston, A. C. Gossard, V. Jaccarino, and Y. Yafet, *Rev. Mod. Phys.* **36**, 170 (1964).

[93] R. J. Noer and W. D. Knight, *Rev. Mod. Phys.* **36**, 177 (1964).

[94] W. A. Hines and W. D. Knight, *Phys. Rev. B* **4**, 893 (1971).

[95] N. J. Curro, T. Caldwell, E. D. Bauer, L. A. Morales, M. J. Graf, Y. Bang, A. V. Balatsky, J. D. Thompson, and J. L. Sarrao, *Nature* **434**, 622 (2005).

[96] V. A. Stenger, C. H. Pennington, D. R. Buffinger, and R. P. Ziebarth, *Phys. Rev. Lett.* **74**, 1649 (1995).

[97] J. A. OSBORN, *Phys. Rev.* **67**, 351 (1945).

[98] M. TINKHAM, *Introduction to Superconductivity*, Dover Publications, Inc., New York, second edition, 2004.

[99] H. ESCHRIG, *Theory of Superconductivity - A Primer*, Lecture Notes, TU Dresden, 2008.

[100] J. WOSNITZA, *Supraleitung*, Lecture Notes, TU Dresden, 2006/2007.

[101] N. BOGOLJUBOV, *Il Nuovo Cimento (1955-1965)* **7**, 794 (1958).

[102] J. VALATIN, *Il Nuovo Cimento (1955-1965)* **7**, 843 (1958).

[103] K. MASUDA and S. KURIHARA, *J. Phys. Soc. Jpn.* **79**, 074710 (2010).

[104] M. NISHIYAMA, Y. INADA, and G.-Q. ZHENG, *Phys. Rev. B* **71**, 220505 (2005).

[105] R. F. KIEFL, W. A. MACFARLANE, K. H. CHOW, S. DUNSIGER, T. L. DUTY, T. M. S. JOHNSTON, J. W. SCHNEIDER, J. SONIER, L. BRARD, R. M. STRONGIN, J. E. FISCHER, and A. B. SMITH, *Phys. Rev. Lett.* **70**, 3987 (1993).

[106] R. W. MORSE and H. V. BOHM, *Phys. Rev.* **108**, 1094 (1957).

[107] K. ASAYAMA, Y. KITAOKA, G. QING ZHENG, and K. ISHIDA, *Progress in Nuclear Magnetic Resonance Spectroscopy* **28**, 221 (1996).

[108] S. OHSUGI, Y. KITAOKA, M. KYOGAKU, K. ISHIDA, K. ASAYAMA, and T. OHTANI, *J. Phys. Soc. Jpn.* **61**, 3054 (1992).

[109] I. GIAEVER and K. MEGERLE, *Phys. Rev.* **122**, 1101 (1961).

[110] C. G. OLSON, R. LIU, A. B. YANG, D. W. LYNCH, A. J. ARKO, R. S. LIST, B. W. VEAL, Y. C. CHANG, P. Z. JIANG, and A. P. PAULIKAS, *Science* **245**, 731 (1989).

[111] S. V. BORISENKO, V. B. ZABOLOTNYY, D. V. EVTUSHINSKY, T. K. KIM, I. V. MOROZOV, A. N. YARESKO, A. A. KORDYUK, G. BEHR, A. VASILIEV, R. FOLLATH, and B. BÜCHNER, *Phys. Rev. Lett.* **105**, 067002 (2010).

[112] R. KHASANOV, D. V. EVTUSHINSKY, A. AMATO, H.-H. KLAUSS, H. LUETKENS, C. NIEDERMAYER, B. BÜCHNER, G. L. SUN, C. T. LIN, J. T. PARK, D. S. INOSOV, and V. HINKOV, *Phys. Rev. Lett.* **102**, 187005 (2009).

[113] M. M. KORSHUNOV and I. EREMIN, *Phys. Rev. B* **78**, 140509 (2008).

[114] T. A. MAIER and D. J. SCALAPINO, *Phys. Rev. B* **78**, 020514 (2008).

[115] T. A. MAIER, S. GRASER, D. J. SCALAPINO, and P. J. HIRSCHFELD, *Phys. Rev. B* **79**, 224510 (2009).

[116] A. D. Christianson, E. A. Goremychkin, R. Osborn, S. Rosenkranz, M. D. Lumsden, C. D. Malliakas, I. S. Todorov, H. Claus, D. Y. Chung, M. G. Kanatzidis, R. I. Bewley, and T. Guidi, *Nature* **456**, 930 (2008).

[117] S. Onari, H. Kontani, and M. Sato, *Phys. Rev. B* **81**, 060504(R) (2010).

[118] S. Lebègue, *Phys. Rev. B* **75**, 035110 (2007).

[119] D. J. Singh and M.-H. Du, *Phys. Rev. Lett.* **100**, 237003 (2008).

[120] L. Boeri, O. V. Dolgov, and A. A. Golubov, *Phys. Rev. Lett.* **101**, 026403 (2008).

[121] I. I. Mazin, D. J. Singh, M. D. Johannes, and M. H. Du, *Phys. Rev. Lett.* **101**, 057003 (2008).

[122] K. Kuroki, S. Onari, R. Arita, H. Usui, Y. Tanaka, H. Kontani, and H. Aoki, *Phys. Rev. Lett.* **101**, 087004 (2008).

[123] D. Singh, *Physica C: Superconductivity* **469**, 418 (2009), Superconductivity in Iron-Pnictides.

[124] D. Lu, M. Yi, S.-K. Mo, J. Analytis, J.-H. Chu, A. Erickson, D. Singh, Z. Hussain, T. Geballe, I. Fisher, and Z.-X. Shen, *Physica C: Superconductivity* **469**, 452 (2009).

[125] E. Dagotto, *Rev. Mod. Phys.* **66**, 763 (1994).

[126] P. A. Lee, N. Nagaosa, and X.-G. Wen, *Rev. Mod. Phys.* **78**, 17 (2006).

[127] C. Day, *Physics Today* **62**, 36 (2009).

[128] C. Chu, F. Chen, M. Gooch, A. Guloy, B. Lorenz, B. Lv, K. Sasmal, Z. Tang, J. Tapp, and Y. Xue, *Physica C: Superconductivity* **469**, 326 (2009).

[129] I. Morozov, A. Boltalin, O. Volkova, A. Vasiliev, O. Kataeva, U. Stockert, M. Abdel-Hafiez, D. Bombor, A. Bachmann, L. Harnagea, M. Fuchs, H.-J. Grafe, G. Behr, R. Klingeler, S. Borisenko, C. Hess, S. Wurmehl, and B. Büchner, *Crystal Growth & Design* **10**, 4428 (2010).

[130] H. Kito, H. Eisaki, and A. Iyo, *J. Phys. Soc. Jpn.* **77**, 063707 (2008).

[131] A. S. Sefat, A. Huq, M. A. McGuire, R. Jin, B. C. Sales, D. Mandrus, L. M. D. Cranswick, P. W. Stephens, and K. H. Stone, *Phys. Rev. B* **78**, 104505 (2008).

[132] A. S. Sefat, R. Jin, M. A. McGuire, B. C. Sales, D. J. Singh, and D. Mandrus, *Phys. Rev. Lett.* **101**, 117004 (2008).

[133] A. Damascelli, Z. Hussain, and Z.-X. Shen, *Rev. Mod. Phys.* **75**, 473 (2003).

[134] O. Fischer, M. Kugler, I. Maggio-Aprile, C. Berthod, and C. Renner, *Rev. Mod. Phys.* **79**, 353 (2007).

[135] H. Luetkens, H.-H. Klauss, M. Kraken, F. J. Litterst, T. Dellmann, R. Klingeler, C. Hess, R. Khasanov, A. Amato, C. Baines, M. Kosmala, O. J. Schumann, M. Braden, J. Hamann-Borrero, N. Leps, A. Kondrat, G. Behr, J. Werner, and B. Büchner, *Nature Mater.* **8**, 305 (2009).

[136] Jun Zhao, Q. Huang, C. de la Cruz, Shiliang Li, J. W. Lynn, Y. Chen, M. A. Green, G. F. Chen, G. Li, Z. Li, J. L. Luo, N. L. Wang, and Pengcheng Dai, *Nature Mater.* **7**, 953 (2008).

[137] A. J. Drew, Ch. Niedermayer, P. J. Baker, F. L. Pratt, S. J. Blundell, T. Lancaster, R. H. Liu, G. Wu, X. H. Chen, I. Watanabe, V. K. Malik, A. Dubroka, M. Rössle, K. W. Kim, C. Baines, and C. Bernhard, *Nature Mater.* **8**, 310 (2009).

[138] S. Sanna, R. De Renzi, G. Lamura, C. Ferdeghini, A. Palenzona, M. Putti, M. Tropeano, and T. Shiroka, *Phys. Rev. B* **80**, 052503 (2009).

[139] S. Sanna, R. De Renzi, T. Shiroka, G. Lamura, G. Prando, P. Carretta, M. Putti, A. Martinelli, M. R. Cimberle, M. Tropeano, and A. Palenzona, *Phys. Rev. B* **82**, 060508(R) (2010).

[140] D. H. Ryan, J. M. Cadogan, C. Ritter, F. Canepa, A. Palenzona, and M. Putti, *Phys. Rev. B* **80**, 220503 (2009).

[141] H. Maeter, H. Luetkens, Y. G. Pashkevich, A. Kwadrin, R. Khasanov, A. Amato, A. A. Gusev, K. V. Lamonova, D. A. Chervinskii, R. Klingeler, C. Hess, G. Behr, B. Büchner, and H.-H. Klauss, *Phys. Rev. B* **80**, 094524 (2009).

[142] G. Prando, P. Carretta, A. Rigamonti, S. Sanna, A. Palenzona, M. Putti, and M. Tropeano, *Phys. Rev. B* **81**, 100508(R) (2010).

[143] A. Alfonsov, F. Murányi, V. Kataev, G. Lang, N. Leps, L. Wang, R. Klingeler, A. Kondrat, C. Hess, S. Wurmehl, A. Köhler, G. Behr, S. Hampel, M. Deutschmann, S. Katrych, N. D. Zhigadlo, Z. Bukowski, J. Karpinski, and B. Büchner, *Phys. Rev. B* **83**, 094526 (2011).

[144] B. Mitrović and K. V. Samokhin, *Phys. Rev. B* **74**, 144510 (2006).

[145] R. Akis and J. P. Carbotte, *Solid State Communications* **78**, 393 (1991).

[146] M. Sigrist and K. Ueda, *Rev. Mod. Phys.* **63**, 239 (1991).

[147] K. Matano, Z. A. Ren, X. L. Dong, L. L. Sun, Z. X. Zhao, and G. q. Zheng, *Europhys. Lett.* **83**, 57001 (2008).

[148] H. Mukuda, N. Terasaki, H. Kinouchi, M. Yashima, Y. Kitaoka, S. Suzuki, S. Miyasaka, S. Tajima, K. Miyazawa, P. Shirage, H. Kito, H. Eisaki, and A. Iyo, *J. Phys. Soc. Jpn.* **77**, 093704 (2008).

[149] Y. Nakai, K. Ishida, Y. Kamihara, M. Hirano, and H. HosonogGg, *J. Phys. Soc. Jpn.* **77**, 073701 (2008).

[150] H. Fukazawa, T. Yamazaki, K. Kondo, Y. Kohori, N. Takeshita, P. M. Shirage, K. Kihou, K. Miyazawa, H. Kito, H. Eisaki, and A. Iyo, *J. Phys. Soc. Jpn.* **78**, 033704 (2009).

[151] K. Hashimoto, T. Shibauchi, T. Kato, K. Ikada, R. Okazaki, H. Shishido, M. Ishikado, H. Kito, A. Iyo, H. Eisaki, S. Shamoto, and Y. Matsuda, *Phys. Rev. Lett.* **102**, 017002 (2009).

[152] L. Malone, J. D. Fletcher, A. Serafin, A. Carrington, N. D. Zhigadlo, Z. Bukowski, S. Katrych, and J. Karpinski, *Phys. Rev. B* **79**, 140501(R) (2009).

[153] H. Ding, P. Richard, K. Nakayama, K. Sugawara, T. Arakane, Y. Sekiba, A. Takayama, S. Souma, T. Sato, T. Takahashi, Z. Wang, X. Dai, Z. Fang, G. F. Chen, J. L. Luo, and N. L. Wang, *Europhys. Lett.* **83**, 47001 (2008).

[154] K. Hashimoto, T. Shibauchi, S. Kasahara, K. Ikada, S. Tonegawa, T. Kato, R. Okazaki, C. J. van der Beek, M. Konczykowski, H. Takeya, K. Hirata, T. Terashima, and Y. Matsuda, *Phys. Rev. Lett.* **102**, 207001 (2009).

[155] X. G. Luo, M. A. Tanatar, J.-P. Reid, H. Shakeripour, N. Doiron-Leyraud, N. Ni, S. L. Bud'ko, P. C. Canfield, H. Luo, Z. Wang, H.-H. Wen, R. Prozorov, and L. Taillefer, *Phys. Rev. B* **80**, 140503(R) (2009).

[156] T. Kondo, A. F. Santander-Syro, O. Copie, C. Liu, M. E. Tillman, E. D. Mun, J. Schmalian, S. L. Bud'ko, M. A. Tanatar, P. C. Canfield, and A. Kaminski, *Phys. Rev. Lett.* **101**, 147003 (2008).

[157] R. S. Gonnelli, D. Daghero, M. Tortello, G. A. Ummarino, V. A. Stepanov, J. S. Kim, and R. K. Kremer, *Phys. Rev. B* **79**, 184526 (2009).

[158] K. A. Yates, L. F. Cohen, Z.-A. Ren, J. Yang, W. Lu, X.-L. Dong, and Z.-X. Zhao, *Superconductor Science and Technology* **21**, 092003 (2008).

[159] Y. Y. Chen, Z. Tesanovic, R. H. Liu, X. H. Chen, and C. L. Chien, *Nature* **453**, 1224 (2008).

[160] S. Chi, A. Schneidewind, J. Zhao, L. W. Harriger, L. Li, Y. Luo, G. Cao, Z. Xu, M. Loewenhaupt, J. Hu, and P. Dai, *Phys. Rev. Lett.* **102**, 107006 (2009).

[161] M. D. Lumsden, A. D. Christianson, D. Parshall, M. B. Stone, S. E. Nagler, G. J. MacDougall, H. A. Mook, K. Lokshin, T. Egami, D. L. Abernathy, E. A. Goremychkin, R. Osborn, M. A. McGuire, A. S. Sefat, R. Jin, B. C. Sales, and D. Mandrus, *Phys. Rev. Lett.* **102**, 107005 (2009).

[162] Y. Kobayashi, A. Kawabata, S. C. Lee, T. Moyoshi, and M. Sato, *J. Phys. Soc. Jpn.* **78**, 073704 (2009).

[163] D. V. Evtushinsky, D. S. Inosov, V. B. Zabolotnyy, A. Koitzsch, M. Knupfer, B. Büchner, M. S. Viazovska, G. L. Sun, V. Hinkov, A. V. Boris, C. T. Lin, B. Keimer, A. Varykhalov, A. A. Kordyuk, and S. V. Borisenko, *Phys. Rev. B* **79**, 054517 (2009).

[164] A. Kawabata, S. C. Lee, T. Moyoshi, Y. Kobayashi, and M. Sato, *J. Phys. Soc. Jpn.* **77**, Supplement C, 147 (2008).

[165] S. Kawasaki, K. Shimada, G. F. Chen, J. L. Luo, N. L. Wang, and Guo-qing Zheng, *Phys. Rev. B* **78**, 220506(R) (2008).

[166] F. Ning, K. Ahilan, T. Imai, A. S. Sefat, R. Jin, M. A. McGuire, B. C. Sales, and D. Mandrus, *J. Phys. Soc. Jpn.* **77**, 103705 (2008).

[167] N. Terasaki, H. Mukuda, M. Yashima, Y. Kitaoka, K. Miyazawa, P. M. Shirage, H. Kito, H. Eisaki, and A. Iyo, *J. Phys. Soc. Jpn.* **78**, 013701 (2009).

[168] Y. Nakai, S. Kitagawa, K. Ishida, Y. Kamihara, M. Hirano, and H. Hosonog, *New J. Phys.* **11**, 045004 (2009).

[169] H. Fukazawa, Y. Yamada, K. Kondo, T. Saito, Y. Kohori, K. Kuga, Y. Matsumoto, S. Nakatsuji, H. Kito, P. M. Shirage, K. Kihou, N. Takeshita, Chul-Ho Lee, A. Iyo, and H. Eisaki, *J. Phys. Soc. Jpn.* **78**, 083712 (2009).

[170] K. Matano, Z. Li, G. L. Sun, D. L. Sun, C. T. Lin, M. Ichioka, and G.-q. Zheng, *Europhys. Lett.* **87**, 27012 (2009).

[171] M. Yashima, H. Nishimura, H. Mukuda, Y. Kitaoka, K. Miyazawa, P. M. Shirage, K. Kihou, H. Kito, H. Eisaki, and A. Iyo, *J. Phys. Soc. Jpn.* **78**, 103702 (2009).

[172] S. W. Zhang, L. Ma, Y. D. Hou, J. Zhang, T.-L. Xia, G. F. Chen, J. P. Hu, G. M. Luke, and W. Yu, *Phys. Rev. B* **81**, 012503 (2010).

[173] M. Sato, Y. Kobayashi, S. C. Lee, H. Takahashi, E. Satomi, and Y. Miura, *J. Phys. Soc. Jpn.* **79**, 014710 (2010).

[174] T. Tabuchi, Z. Li, T. Oka, G. F. Chen, S. Kawasaki, J. L. Luo, N. L. Wang, and G.-q. Zheng, *Phys. Rev. B* **81**, 140509 (2010).

[175] A. V. Chubukov, D. V. Efremov, and I. Eremin, *Phys. Rev. B* **78**, 134512 (2008).

[176] D. Parker, O. V. Dolgov, M. M. Korshunov, A. A. Golubov, and I. I. Mazin, *Phys. Rev. B* **78**, 134524 (2008).

[177] Y. Bang, H.-Y. Choi, and H. Won, *Phys. Rev. B* **79**, 054529 (2009).

[178] B. Muschler, W. Prestel, R. Hackl, T. P. Devereaux, J. G. Analytis, J.-H. Chu, and I. R. Fisher, *Phys. Rev. B* **80**, 180510(R) (2009).

[179] A. Kondrat, J. E. Hamann-Borrero, N. Leps, M. Kosmala, O. Schumann, A. Köhler, J. Werner, G. Behr, M. Braden, R. Klingeler, B. Büchner, and C. Hess, *Eur. Phys. J. B* **70**, 461 (2009).

[180] C. Hess, A. Kondrat, A. Narduzzo, J. E. Hamann-Borrero, R. Klingeler, J. Werner, G. Behr, and B. Büchner, *Europhys. Lett.* **87**, 17005 (2009).

[181] R. Klingeler, N. Leps, I. Hellmann, A. Popa, U. Stockert, C. Hess, V. Kataev, H.-J. Grafe, F. Hammerath, G. Lang, S. Wurmehl, G. Behr, L. Harnagea, S. Singh, and B. Büchner, *Phys. Rev. B* **81**, 024506 (2010).

[182] L. Wang, U. Köhler, N. Leps, A. Kondrat, M. Nale, A. Gasparini, A. de Visser, G. Behr, C. Hess, R. Klingeler, and B. Büchner, *Phys. Rev. B* **80**, 094512 (2009).

[183] H.-H. Klauss, H. Luetkens, R. Klingeler, C. Hess, F. J. Litterst, M. Kraken, M. M. Korshunov, I. Eremin, S.-L. Drechsler, R. Khasanov, A. Amato, J. Hamann-Borrero, N. Leps, A. Kondrat, G. Behr, J. Werner, and B. Büchner, *Phys. Rev. Lett.* **101**, 077005 (2008).

[184] M. A. McGuire, A. D. Christianson, A. S. Sefat, B. C. Sales, M. D. Lumsden, R. Jin, E. A. Payzant, D. Mandrus, Y. Luan, V. Keppens, V. Varadarajan, J. W. Brill, R. P. Hermann, M. T. Sougrati, F. Grandjean, and G. J. Long, *Phys. Rev. B* **78**, 094517 (2008).

[185] S. Kitao, Y. Kobayashi, S. Higashitaniguchi, M. Saito, Y. Kamihara, M. Hirano, T. Mitsui, H. Hosono, and M. Seto, *J. Phys. Soc. Jpn.* **77**, 103706 (2008).

[186] C. de la Cruz, Q. Huang, J. W. Lynn, Jiying Li, W. Ratcliff II, J. L. Zarestky, H. A. Mook, G. F. Chen, J. L. Luo, N. L. Wang, and Pengcheng Dai, *Nature* **453**, 899 (2008).

[187] J. Dong, H. J. Zhang, G. Xu, Z. Li, G. Li, W. Z. Hu, D. Wu, G. F. Chen, X. Dai, J. L. Luo, Z. Fang, and N. L. Wang, *Europhys. Lett.* **83**, 27006 (2008).

[188] T. Yildirim, *Phys. Rev. Lett.* **101**, 057010 (2008).

[189] C. Cao, P. J. Hirschfeld, and H.-P. Cheng, *Phys. Rev. B* **77**, 220506 (2008).

[190] F. Ma, Z.-Y. Lu, and T. Xiang, *Phys. Rev. B* **78**, 224517 (2008).

[191] I. I. MAZIN and M. JOHANNES, *Nature Physics* **5**, 141 (2009).

[192] N. QURESHI, Y. DREES, J. WERNER, S. WURMEHL, C. HESS, R. KLINGELER, B. BÜCHNER, M. T. FERNÁNDEZ-DÍAZ, and M. BRADEN, *Phys. Rev. B* **82**, 184521 (2010).

[193] H.-J. GRAFE, G. LANG, F. HAMMERATH, D. PAAR, K. MANTHEY, K. KOCH, H. ROSNER, N. J. CURRO, G. BEHR, J. WERNER, N. LEPS, R. KLINGELER, H.-H. KLAUSS, F. J. LITTERST, and B. BÜCHNER, *New J. Phys.* **11**, 035002 (2009).

[194] J. ZHAO, Q. HUANG, C. DE LA CRUZ, J. W. LYNN, M. D. LUMSDEN, Z. A. REN, J. YANG, X. SHEN, X. DONG, Z. ZHAO, and P. DAI, *Phys. Rev. B* **78**, 132504 (2008).

[195] Y. QIU, W. BAO, Q. HUANG, T. YILDIRIM, J. M. SIMMONS, M. A. GREEN, J. W. LYNN, Y. C. GASPAROVIC, J. LI, T. WU, G. WU, and X. H. CHEN, *Phys. Rev. Lett.* **101**, 257002 (2008).

[196] Y. CHEN, J. W. LYNN, J. LI, G. LI, G. F. CHEN, J. L. LUO, N. L. WANG, P. DAI, C. DELA CRUZ, and H. A. MOOK, *Phys. Rev. B* **78**, 064515 (2008).

[197] S. A. J. KIMBER, D. N. ARGYRIOU, F. YOKAICHIYA, K. HABICHT, S. GERISCHER, T. HANSEN, T. CHATTERJI, R. KLINGELER, C. HESS, G. BEHR, A. KONDRAT, and B. BÜCHNER, *Phys. Rev. B* **78**, 140503 (2008).

[198] J. W. LYNN and P. DAI, *Physica C: Superconductivity* **469**, 469 (2009).

[199] Q. HUANG, Y. QIU, W. BAO, M. A. GREEN, J. W. LYNN, Y. C. GASPAROVIC, T. WU, G. WU, and X. H. CHEN, *Phys. Rev. Lett.* **101**, 257003 (2008).

[200] A. I. GOLDMAN, D. N. ARGYRIOU, B. OULADDIAF, T. CHATTERJI, A. KREYSSIG, S. NANDI, N. NI, S. L. BUD'KO, P. C. CANFIELD, and R. J. MCQUEENEY, *Phys. Rev. B* **78**, 100506 (2008).

[201] J. ZHAO, W. RATCLIFF, J. W. LYNN, G. F. CHEN, J. L. LUO, N. L. WANG, J. HU, and P. DAI, *Phys. Rev. B* **78**, 140504 (2008).

[202] K. KANEKO, A. HOSER, N. CAROCA-CANALES, A. JESCHE, C. KRELLNER, O. STOCKERT, and C. GEIBEL, *Phys. Rev. B* **78**, 212502 (2008).

[203] A. JESCHE, N. CAROCA-CANALES, H. ROSNER, H. BORRMANN, A. ORMECI, D. KASINATHAN, H. H. KLAUSS, H. LUETKENS, R. KHASANOV, A. AMATO, A. HOSER, K. KANEKO, C. KRELLNER, and C. GEIBEL, *Phys. Rev. B* **78**, 180504 (2008).

[204] L. X. YANG, B. P. XIE, Y. ZHANG, C. HE, Q. Q. GE, X. F. WANG, X. H. CHEN, M. ARITA, J. JIANG, K. SHIMADA, M. TANIGUCHI, I. VOBORNIK, G. ROSSI, J. P. HU, D. H. LU, Z. X. SHEN, Z. Y. LU, and D. L. FENG, *Phys. Rev. B* **82**, 104519 (2010).

[205] C. Liu, Y. Lee, A. D. Palczewski, J.-Q. Yan, T. Kondo, B. N. Harmon, R. W. McCallum, T. A. Lograsso, and A. Kaminski, *Phys. Rev. B* **82**, 075135 (2010).

[206] T. Sato, S. Souma, K. Nakayama, K. Terashima, K. Sugawara, T. Takahashi, Y. Kamihara, M. Hirano, and H. Hosono, *J. Phys. Soc. Jpn.* **77**, 063708 (2008).

[207] U. Stockert, M. Abdel Hafiez, D. V. Evtushinsky, V. B. Zabolotnyy, A. U. B. Wolter, S. Wurmehl, I. Morozov, R. Klingeler, S. V. Borisenko, and B. Büchner, *arXiv:1011.4246v1* .

[208] D. J. Singh, *Phys. Rev. B* **78**, 094511 (2008).

[209] Y.-F. Li and B.-G. Liu, *Eur. Phys. J. B* **72**, 153 (2009).

[210] R. A. Jishi and H. M. Alyahyaei, *Advances in Condensed Matter Physics* **2010**, 804343 (2010).

[211] W. L. McMillan, *Phys. Rev.* **167**, 331 (1968).

[212] P. B. Allen and R. C. Dynes, *Phys. Rev. B* **12**, 905 (1975).

[213] T. Hänke, S. Sykora, R. Schlegel, D. Baumann, L. Harnagea, S. Wurmehl, M. Daghofer, B. Büchner, J. van den Brink, and C. Hess, *arXiv:1106.4217* (2011).

[214] P. M. R. Brydon, M. Daghofer, C. Timm, and J. van den Brink, *Phys. Rev. B* **83**, 060501 (2011).

[215] N. W. Ashcroft and D. N. Mermin, *Festkörperphysik*, Oldenbourg, München, third edition, 2007.

[216] S. Aswartham, G. Behr, L. Harnagea, D. Bombor, A. Bachmann, I. V. Morozov, V. B. Zabolotnyy, A. A. Kordyuk, T. Kim, D. V. Evtushinsky, S. V. Borisenko, A. U. B. Wolter, C. Hess, S. Wurmehl, and B. Büchner, *Phys. Rev. B* **84**, 054534 (2011).

[217] X. Zhu, H. Yang, L. Fang, G. Mu, and H.-H. Wen, *Superconductor Science and Technology* **21**, 105001 (2008).

[218] H. Luetkens, H.-H. Klauss, R. Khasanov, A. Amato, R. Klingeler, I. Hellmann, N. Leps, A. Kondrat, C. Hess, A. Köhler, G. Behr, J. Werner, and B. Büchner, *Phys. Rev. Lett.* **101**, 097009 (2008).

[219] G. Fuchs, S.-L. Drechsler, N. Kozlova, G. Behr, A. Köhler, J. Werner, K. Nenkov, R. Klingeler, J. Hamann-Borrero, C. Hess, A. Kondrat, M. Grobosch, A. Narduzzo, M. Knupfer, J. Freudenberger, B. Büchner, and L. Schultz, *Phys. Rev. Lett.* **101**, 237003 (2008).

[220] B. C. CHANG, C. H. HSU, Y. Y. HSU, Z. WEI, K. Q. RUAN, X. G. LI, and H. C. KU, *Europhys. Lett.* **84**, 67014 (2008).

[221] A. OLARIU, F. RULLIER-ALBENQUE, D. COLSON, and A. FORGET, *Phys. Rev. B* **83**, 054518 (2011).

[222] S. R. SAHA, N. P. BUTCH, K. KIRSHENBAUM, and J. PAGLIONE, *Phys. Rev. B* **79**, 224519 (2009).

[223] O. HEYER, T. LORENZ, V. B. ZABOLOTNYY, D. V. EVTUSHINSKY, S. V. BORISENKO, I. MOROZOV, L. HARNAGEA, S. WURMEHL, C. HESS, and B. BÜCHNER, *Phys. Rev. B* **84**, 064512 (2011).

[224] N. J. CURRO, A. P. DIOGUARDI, N. APROBERTS-WARREN, A. C. SHOCKLEY, and P. KLAVINS, *New J. Phys.* **11**, 075004 (2009).

[225] L. MA, J. ZHANG, G. F. CHEN, and W. YU, *Phys. Rev. B* **82**, 180501 (2010).

[226] K. AHILAN, F. L. NING, T. IMAI, A. S. SEFAT, R. JIN, M. A. MCGUIRE, B. C. SALES, and D. MANDRUS, *Phys. Rev. B* **78**, 100501(R) (2008).

[227] T. IMAI, K. AHILAN, F. NING, M. A. MCGUIRE, A. S. SEFAT, R. JIN, B. C. SALES, and D. MANDRUS, *J. Phys. Soc. Jpn.* **77SC**, 47 (2008).

[228] K. KITAGAWA, N. KATAYAMA, K. OHGUSHI, M. YOSHIDA, and M. TAKIGAWA, *J. Phys. Soc. Jpn.* **77**, 114709 (2008).

[229] F. NING, K. AHILAN, T. IMAI, A. S. SEFAT, R. JIN, M. A. MCGUIRE, B. C. SALES, and D. MANDRUS, *J. Phys. Soc. Jpn.* **78**, 013711 (2009).

[230] F. L. NING, K. AHILAN, T. IMAI, A. S. SEFAT, M. A. MCGUIRE, B. C. SALES, D. MANDRUS, P. CHENG, B. SHEN, and H.-H. WEN, *Phys. Rev. Lett.* **104**, 037001 (2010).

[231] N. J. CURRO, T. IMAI, C. P. SLICHTER, and B. DABROWSKI, *Phys. Rev. B* **56**, 877 (1997).

[232] D. MIHAILOVIC, V. V. KABANOV, K. ŽAGAR, and J. DEMSAR, *Phys. Rev. B* **60**, R6995 (1999).

[233] T. IMAI, A. W. HUNT, K. R. THURBER, and F. C. CHOU, *Phys. Rev. Lett.* **81**, 3006 (1998).

[234] A. M. CLOGSTON and V. JACCARINO, *Phys. Rev.* **121**, 1357 (1961).

[235] A. M. CLOGSTON, V. JACCARINO, and Y. YAFET, *Phys. Rev.* **134**, A650 (1964).

[236] K. HAULE, J. H. SHIM, and G. KOTLIAR, *Phys. Rev. Lett.* **100**, 226402 (2008).

[237] H. MUKUDA, N. TERASAKI, N. TAMURA, H. KINOUCHI, M. YASHIMA, Y. KITAOKA, K. MIYAZAWA, P. M. SHIRAGE, S. SUZUKI, S. MIYASAKA, S. TAJIMA, H. KITO, H. EISAKI, and A. IYO, *J. Phys. Soc. Jpn.* **78**, 084717 (2009).

[238] Q. Huang, J. Zhao, J. W. Lynn, G. F. Chen, J. L. Luo, N. L. Wang, and P. Dai, *Phys. Rev. B* **78**, 054529 (2008).

[239] Q. Si and E. Abrahams, *Phys. Rev. Lett.* **101**, 076401 (2008).

[240] J. Wu, P. Phillips, and A. H. Castro Neto, *Phys. Rev. Lett.* **101**, 126401 (2008).

[241] G. M. Zhang, Y. H. Su, Z. Y. Lu, Z. Y. Weng, D. H. Lee, and T. Xiang, *Europhys. Lett.* **86**, 37006 (2009).

[242] M. M. Korshunov and I. Eremin, *Europhys. Lett.* **83**, 67003 (2008).

[243] M. M. Korshunov, I. Eremin, D. V. Efremov, D. L. Maslov, and A. V. Chubukov, *Phys. Rev. Lett.* **102**, 236403 (2009).

[244] M. Berciu, I. Elfimov, and G. A. Sawatzky, *Phys. Rev. B* **79**, 214507 (2009).

[245] G. A. Sawatzky, I. S. Elfimov, J. van den Brink, and J. Zaanen, *Europhys. Lett.* **86**, 17006 (2009).

[246] T. Timusk and B. Statt, *Reports on Progress in Physics* **62**, 61 (1999).

[247] D. Paar, H.-J. Grafe, G. Lang, F. Hammerath, K. Manthey, G. Behr, J. Werner, and B. Büchner, *Physica C: Superconductivity* **470**, S468 (2010), Proceedings of the 9th International Conference on Materials and Mechanisms of Superconductivity.

[248] K. Tatsumi, N. Fujiwara, H. Okada, H. Takahashi, Y. Kamihara, M. Hirano, and H. Hosono, *J. Phys. Soc. Jpn.* **78**, 023709 (2009).

[249] R. Walstedt and W. Warren, *Applied Magnetic Resonance* **3**, 469 (1992).

[250] N. J. Curro and Morales, *MRS Proceedings* **802**, DD2.4 (2003).

[251] S.-H. Baek, N. J. Curro, T. Klimczuk, E. D. Bauer, F. Ronning, and J. D. Thompson, *Phys. Rev. B* **79**, 052504 (2009).

[252] Y. Nakai, T. Iye, S. Kitagawa, K. Ishida, H. Ikeda, S. Kasahara, H. Shishido, T. Shibauchi, Y. Matsuda, and T. Terashima, *Phys. Rev. Lett.* **105**, 107003 (2010).

[253] T. Imai, K. Ahilan, F. L. Ning, T. M. McQueen, and R. J. Cava, *Phys. Rev. Lett.* **102**, 177005 (2009).

[254] D. A. Torchetti, M. Fu, D. C. Christensen, K. J. Nelson, T. Imai, H. C. Lei, and C. Petrovic, *Phys. Rev. B* **83**, 104508 (2011).

[255] X. F. Wang, T. Wu, G. Wu, R. H. Liu, H. Chen, Y. L. Xie, and X. H. Chen, *New J. Phys.* **11**, 045003 (2009).

[256] JIUN-HAW CHU, JAMES G. ANALYTIS, CHRIS KUCHARCZYK, and IAN R. FISHER, *Phys. Rev. B* **79**, 014506 (2009).

[257] C. LESTER, J.-H. CHU, J. G. ANALYTIS, S. C. CAPELLI, A. S. ERICKSON, C. L. CONDRON, M. F. TONEY, I. R. FISHER, and S. M. HAYDEN, *Phys. Rev. B* **79**, 144523 (2009).

[258] S. ASWARTHAM, C. NACKE, G. FRIEMEL, N. LEPS, S. WURMEHL, N. WIZENT, C. HESS, R. KLINGELER, G. BEHR, S. SINGH, and B. BÜCHNER, *Journal of Crystal Growth* **314**, 341 (2011).

[259] K. UEDA and T. MORIYA, *J. Phys. Soc. Jpn.* **38**, 32 (1975).

[260] T. MORIYA, *Spin Fluctuations in Itinerant Electron Magnetism*, volume 56 of *Springer Series in Solid-State Sciences*, Springer, Berlin, 1985.

[261] S. OHSUGI, Y. KITAOKA, K. ISHIDA, and K. ASAYAMA, *J. Phys. Soc. Jpn.* **60**, 2351 (1991).

[262] B. J. SUH, P. C. HAMMEL, Y. YOSHINARI, J. D. THOMPSON, J. L. SARRAO, and Z. FISK, *Phys. Rev. Lett.* **81**, 2791 (1998).

[263] F. C. CHOU, F. BORSA, J. H. CHO, D. C. JOHNSTON, A. LASCIALFARI, D. R. TORGESON, and J. ZIOLO, *Phys. Rev. Lett.* **71**, 2323 (1993).

[264] L. FANG, H. LUO, P. CHENG, Z. WANG, Y. JIA, G. MU, B. SHEN, I. I. MAZIN, L. SHAN, C. REN, and H.-H. WEN, *Phys. Rev. B* **80**, 140508 (2009).

[265] A. S. SEFAT, M. A. MCGUIRE, B. C. SALES, R. JIN, J. Y. HOWE, and D. MANDRUS, *Phys. Rev. B* **77**, 174503 (2008).

[266] M. BANKAY, M. MALI, J. ROOS, and D. BRINKMANN, *Phys. Rev. B* **50**, 6416 (1994).

[267] N. DOIRON-LEYRAUD, P. AUBAN-SENZIER, S. RENÉ DE COTRET, C. BOURBONNAIS, D. JÉROME, K. BECHGAARD, and L. TAILLEFER, *Phys. Rev. B* **80**, 214531 (2009).

[268] S. KASAHARA, T. SHIBAUCHI, K. HASHIMOTO, K. IKADA, S. TONEGAWA, R. OKAZAKI, H. SHISHIDO, H. IKEDA, H. TAKEYA, K. HIRATA, T. TERASHIMA, and Y. MATSUDA, *Phys. Rev. B* **81**, 184519 (2010).

[269] K. YOSHIMURA, T. SHIMIZU, M. TAKIGAWA, H. YASUOKA, and Y. NAKAMURA, *J. Phys. Soc. Jpn.* **53**, 503 (1984).

[270] H. ALLOUL and L. MIHALY, *Phys. Rev. Lett.* **48**, 1420 (1982).

[271] G. QING ZHENG, Y. KITAOKA, K. ASAYAMA, Y. KODAMA, and Y. YAMADA, *Physica C: Superconductivity* **193**, 154 (1992).

[272] H. ALLOUL, A. MAHAJAN, H. CASALTA, and O. KLEIN, *Phys. Rev. Lett.* **70**, 1171 (1993).

[273] M. Horvatić, C. Berthier, Y. Berthier, P. Ségransan, P. Butaud, W. G. Clark, J. A. Gillet, and J. Y. Henry, *Phys. Rev. B* **48**, 13848 (1993).

[274] Y. Yoshinari, H. Yasuoka, Y. Ueda, K. ichi Koga, and K. Kosuge, *J. Phys. Soc. Jpn.* **59**, 3698 (1990).

[275] A. Kawabata, S. C. Lee, T. Moyoshi, Y. Kobayashi, and M. Sato, *J. Phys. Soc. Jpn.* **77**, 103704 (2008).

[276] H. Mukuda, N. Terasaki, M. Yashima, H. Nishimura, Y. Kitaoka, and A. Iyo, *Physica C: Superconductivity* **469**, 559 (2009).

[277] Y. Nakai, T. Iye, S. Kitagawa, K. Ishida, S. Kasahara, T. Shibauchi, Y. Matsuda, and T. Terashima, *Phys. Rev. B* **81**, 020503(R) (2010).

[278] S. Oh, A. M. Mounce, S. Mukhopadhyay, W. P. Halperin, A. B. Vorontsov, S. L. Bud'ko, P. C. Canfield, Y. Furukawa, A. P. Reyes, and P. L. Kuhns, *Phys. Rev. B* **83**, 214501 (2011).

[279] F. Hammerath, S.-L. Drechsler, H.-J. Grafe, G. Lang, G. Fuchs, G. Behr, I. Eremin, M. M. Korshunov, and B. Büchner, *Phys. Rev. B* **81**, 140504(R) (2010).

[280] K. Hashimoto, M. Yamashita, S. Kasahara, Y. Senshu, N. Nakata, S. Tonegawa, K. Ikada, A. Serafin, A. Carrington, T. Terashima, H. Ikeda, T. Shibauchi, and Y. Matsuda, *Phys. Rev. B* **81**, 220501 (2010).

[281] J. D. Fletcher, A. Serafin, L. Malone, J. G. Analytis, J.-H. Chu, A. S. Erickson, I. R. Fisher, and A. Carrington, *Phys. Rev. Lett.* **102**, 147001 (2009).

[282] C. W. Hicks, T. M. Lippman, M. E. Huber, J. G. Analytis, J.-H. Chu, A. S. Erickson, I. R. Fisher, and K. A. Moler, *Phys. Rev. Lett.* **103**, 127003 (2009).

[283] C. Martin, H. Kim, R. T. Gordon, N. Ni, V. G. Kogan, S. L. Bud'ko, P. C. Canfield, M. A. Tanatar, and R. Prozorov, *Phys. Rev. B* **81**, 060505 (2010).

[284] K. Kuroki, H. Usui, S. Onari, R. Arita, and H. Aoki, *Phys. Rev. B* **79**, 224511 (2009).

[285] V. Vildosola, L. Pourovskii, R. Arita, S. Biermann, and A. Georges, *Phys. Rev. B* **78**, 064518 (2008).

[286] H. Kontani and S. Onari, *Phys. Rev. Lett.* **104**, 157001 (2010).

[287] P. Anderson, *Journal of Physics and Chemistry of Solids* **11**, 26 (1959).

[288] A. A. Golubov and I. I. Mazin, *Phys. Rev. B* **55**, 15146 (1997).

[289] Y. Nagai, N. Hayashi, N. Nakai, H. Nakamura, M. Okumura, and M. Machida, *New J. Phys.* **10**, 103026 (2008).

[290] G. Fuchs, S.-L. Drechsler, N. Kozlova, M. Bartkowiak, J. E. Hamann-Borrero, G. Behr, K. Nenkov, H.-H. Klauss, H. Maeter, A. Amato, H. Luetkens, A. Kwadrin, R. Khasanov, J. Freudenberger, A. Köhler, M. Knupfer, E. Arushanov, H. Rosner, B. Büchner, and L. Schultz, *New J. Phys.* **11**, 075007 (2009).

[291] V. Grinenko, K. Kikoin, S.-L. Drechsler, G. Fuchs, K. Nenkov, S. Wurmehl, F. Hammerath, G. Lang, H.-J. Grafe, B. Holzapfel, J. van den Brink, B. Büchner, and L. Schultz, *Phys. Rev. B* **84**, 134516 (2011).

[292] T. Saito, K. Koyama, K. ichi Magishi, and K. Endo, *Journal of Magnetism and Magnetic Materials* **310**, 681 (2007), Proceedings of the 17th International Conference on Magnetism, The International Conference on Magnetism.

[293] E. Ehrenfreund, I. B. Goldberg, and M. Weger, *Solid State Communications* **7**, 1333 (1969).

[294] Y. Kobayashi, E. Satomi, S. C. Lee, and M. Sato, *J. Phys. Soc. Jpn.* **79**, 093709 (2010).

[295] A. V. Chubukov, M. G. Vavilov, and A. B. Vorontsov, *Phys. Rev. B* **80**, 140515(R) (2009).

[296] Y. Yanagi, Y. Yamakawa, and Y. Omacrno, *Phys. Rev. B* **81**, 054518 (2010).

[297] Y. Nakai, K. Ishida, Y. Kamihara, M. Hirano, and H. Hosono, *Phys. Rev. Lett.* **101**, 077006 (2008).

[298] S.-H. Baek, H.-J. Grafe, F. Hammerath, M. Fuchs, C. Rudisch, L. Harnagea, S. Aswartham, S. Wurmehl, J. van den Brink, and B. Büchner, *arXiv:1108.2592* (2011).

[299] S.-H. Baek, H.-J. Grafe, L. Harnagea, S. Singh, S. Wurmehl, and B. Büchner, *Phys. Rev. B* **84**, 094510 (2011).

[300] G. Lang, *private communication* .

[301] K. Kitagawa, N. Katayama, K. Ohgushi, and M. Takigawa, *J. Phys. Soc. Jpn.* **78**, 063706 (2009).

[302] K. Deguchi, M. A. Tanatar, Z. Mao, T. Ishiguro, and Y. Maeno, *J. Phys. Soc. Jpn.* **71**, 2839 (2002).

[303] H. Shimahara, *J. Phys. Soc. Jpn.* **69**, 1966 (2000).

[304] J. Shinagawa, Y. Kurosaki, F. Zhang, C. Parker, S. E. Brown, D. Jérome, J. B. Christensen, and K. Bechgaard, *Phys. Rev. Lett.* **98**, 147002 (2007).

[305] S. M. De Soto, C. P. Slichter, H. H. Wang, U. Geiser, and J. M. Williams, *Phys. Rev. Lett.* **70**, 2956 (1993).

[306] M. Corti, B. J. Suh, F. Tabak, A. Rigamonti, F. Borsa, M. Xu, and B. Dabrowski, *Phys. Rev. B* **54**, 9469 (1996).

[307] J. Christiansen, P. Heubes, R. Keitel, W. Klinger, W. Loeffler, W. Sandner, and W. Witthuhn, *Zeitschrift für Physik B Condensed Matter* **24**, 177 (1976).

[308] B. Gee, C. R. Horne, E. J. Cairns, and J. A. Reimer, *The Journal of Physical Chemistry B* **102**, 10142 (1998).

[309] A. Narath, *Phys. Rev.* **162**, 320 (1967).

[310] A. Suter, M. Mali, J. Roos, and D. Brinkmann, *Journal of Magnetic Resonance* **143**, 266 (2000).

[311] D. C. Johnston, S.-H. Baek, X. Zong, F. Borsa, J. Schmalian, and S. Kondo, *Phys. Rev. Lett.* **95**, 176408 (2005).

[312] P. Vonlanthen, K. B. Tanaka, A. Goto, W. G. Clark, P. Millet, J. Y. Henry, J. L. Gavilano, H. R. Ott, F. Mila, C. Berthier, M. Horvatić, Y. Tokunaga, P. Kuhns, A. P. Reyes, and W. G. Moulton, *Phys. Rev. B* **65**, 214413 (2002).

[313] Y. Shimizu, K. Miyagawa, K. Kanoda, M. Maesato, and G. Saito, *Phys. Rev. B* **73**, 140407 (2006).

[314] J.-E. Weber, C. Kegler, N. Büttgen, H.-A. Krug von Nidda, A. Loidl, and F. Lichtenberg, *Phys. Rev. B* **64**, 235414 (2001).

[315] B. Simovič, P. C. Hammel, M. Hücker, B. Büchner, and A. Revcolevschi, *Phys. Rev. B* **68**, 012415 (2003).

[316] Y. Itoh, T. Machi, N. Koshizuka, M. Murakami, H. Yamagata, and M. Matsumura, *Phys. Rev. B* **69**, 184503 (2004).

[317] H.-J. Grafe, Curro, N.J., Young, B.L., Vyalikh, A., Vavilova, J., Gu, G.D., Hücker, M., and Büchner, B., *Eur. Phys. J. Special Topics* **188**, 89 (2010).

[318] M.-H. Julien, A. Campana, A. Rigamonti, P. Carretta, F. Borsa, P. Kuhns, A. P. Reyes, W. G. Moulton, M. Horvatić, C. Berthier, A. Vietkin, and A. Revcolevschi, *Phys. Rev. B* **63**, 144508 (2001).

[319] A. Morello, F. L. Mettes, O. N. Bakharev, H. B. Brom, L. J. de Jongh, F. Luis, J. F. Fernández, and G. Aromí, *Phys. Rev. B* **73**, 134406 (2006).

[320] V. F. Mitrović, M.-H. Julien, C. de Vaulx, M. Horvatić, C. Berthier, T. Suzuki, and K. Yamada, *Phys. Rev. B* **78**, 014504 (2008).

[321] G. V. M. Williams and S. Krämer, *Phys. Rev. B* **64**, 104506 (2001).

[322] G. V. M. Williams, H. K. Lee, and S. K. Goh, *Phys. Rev. B* **71**, 014515 (2005).

[323] G. V. M. Williams, *Phys. Rev. B* **73**, 064510 (2006).

[324] G. V. M. Williams, *Phys. Rev. B* **76**, 094502 (2007).

[325] A. Krimmel, A. Günther, W. Kraetschmer, H. Dekinger, N. Büttgen, A. Loidl, S. G. Ebbinghaus, E.-W. Scheidt, and W. Scherer, *Phys. Rev. B* **78**, 165126 (2008).

[326] D. C. Johnston, *Phys. Rev. B* **74**, 184430 (2006).

[327] A. Rigamonti, F. Borsa, and P. Carretta, *Reports on Progress in Physics* **61**, 1367 (1998).

[328] K. Ishida, Y. Kitaoka, and K. Asayama, *Solid State Communications* **90**, 563 (1994).

[329] I. B. Goldberg and M. Weger, *J. Phys. Soc. Jpn.* **24**, 1279 (1968).

# Acknowledgement

Für die Hilfe und Unterstützung beim Erstellen meiner Doktorarbeit möchte ich es nicht versäumen, am Ende einigen wichtigen Menschen herzlich zu danken.

Mein erster Dank gilt meinem Doktorvater, Prof. Dr. Bernd Büchner, der mir die Möglichkeit gab, diese Arbeit am Institut für Festkörperforschung schreiben zu können und sie auf internationalen Konferenzen präsentieren zu dürfen. Sein stetes Interesse an meinen Ergebnissen sowie seine Gepflogenheit, Doktoranden vom ersten Tag an in die wissenschaftliche Diskussion mit einzubeziehen, haben mich immer wieder hoch motiviert.

I owe my deepest gratitude to Prof. Nicholas J. Curro, who straightforwardly agreed in reviewing this thesis.

Ein großer Dank gilt auch meinem Betreuer, Dr. Hans-Joachim Grafe, der mir stets bereitwillig, geduldig und gut gelaunt für Diskussionen zur Verfügung stand und mit konstruktiver Kritik, vielen neuen Ideen, Ermutigung und Vertrauen den größten Anteil am Gelingen dieser Arbeit inne hat. Besonders möchte ich ihm auch für die schnelle Korrektur der Dissertation danken.

I am very grateful to Dr. Guillaume Lang and Steven Rodan for their helpful corrections of the thesis within the very short period of time.

Ein besonders herzlicher Dank gilt der gesamten NMR-Gruppe des IFW Dresdens, Dr. Hans-Joachim Grafe, Regina Vogel, Dr. Anja Wolter, Dr. Guillaume Lang, Dr. Seung-Ho Baek, Christian Rudisch, Markus Schäpers, Katarina Manthey, Yannic Utz, Uwe Gräfe und Dr. Eva Maria Brüning, für ihre Hilfe und Unterstützung, für die besonders gute Arbeitsatmosphäre und für die Bereitstellung ihrer Messergebnisse, mit denen meine Ergebnisse erst ein rundes Bild ergeben. Durch die ausgezeichnete fachliche Zusammenarbeit mit euch und durch das, was wir gemeinsam an persönlichem Schicksal erlebt und verarbeitet haben, habe ich schätzen gelernt, was gute Kollegen ausmacht. Danke für die so wichtigen Stunden, in denen es nicht um Physik ging. Ich hoffe, dass wir noch lange in Kontakt bleiben, und wünsche euch und euren Familien von Herzen alles Gute.

Für die zahlreichen kreativen und lebendigen Diskussionen der Ergebnisse, für die glorreiche Idee der "smart impurities" und für seine ansteckende, spitzbübische Begeisterungsfähigkeit für die Physik danke ich Dr. Stefan-Ludwig Drechsler. Auch Prof. Ilya Eremin möchte ich für den wichtigen theoretischen Input danken. Des Weiteren sei Dr. Günter Fuchs, Dr. Christian Hess, Dr. Vadim Grinenko, Prof. Dr. Rüdiger Klingeler und Prof. Dr. Jeroen van den Brink für ergiebige Diskussionen gedankt.

Für die Bereitstellung hochwertiger Proben, die die Basis dieser Arbeit sind, danke ich Dr. Sabine Wurmehl, Saicharan Aswartham, Dr. Luminita Harnagea, Jochen Werner, Dr. Günther Behr, Margitta Deutschmann, Sabine Müller-Litvanyi und Dr. Anke Köhler.

Falk Herold möchte ich für die schnelle und kompetente Hilfe bei der Verbesserung des Probenstabs und für allerlei kleinere Reparaturen danken.

Meinen Mitstreitern, vor allem meinen Büronachbarn Alexey und Ferdinand, sowie Yulieth, Nikolai, Sai, Claudia, Yulia, Nadya, Martin, Frederik, Stefan, Nadja, Norman, Wolfram, Mohammed, Matthias und Franziska danke ich für die gute Gemeinschaft, die motivierende Zusammenarbeit, nette Kaffeepausen, gemütliche Abende und unzählige Diskussionen über physikalische Sachverhalte und sonstige weltbewegende Dinge.

Ein herzlicher Dank geht an meine Freunde. Ob direkt um die Ecke oder im Norden Europas, ob auf dem Tanzboden oder dem Dachboden, ob in DD oder D, euer Zuspruch, eure Unterstützung, eure Ablenkung, eure Geduld und eure guten Wünsche haben mir gerade in der letzten heißen Phase sehr geholfen.

Mein innigster Dank gilt meiner Familie. Habt Dank für euer Vertrauen, eure Unterstützung und eure Liebe. Ihr habt es vermocht, mir gleichzeitig den Rücken frei zu halten und ihn mir zu stärken. Juri, gracias por tu amor, tu paciencia y tu comprensión.

Franziska Hammerath

Printed by Publishers' Graphics LLC